THE PARI DIALOGUES volume 1

Essays in Science, Religion, Society and the Arts

THE PARI DIALOGUES volume 1

Essays in Science, Religion, Society and the Arts

Edited by F. David Peat

Pari Publishing

First published in Italy in 2007 by Pari Publishing, Sas.

Reprinted in 2008.

Printed and bound in Canada.

A catalogue record for this book is available from the British Library.

ISBN 978-88-901960-5-8

Book design and cover by Andrea Barbieri

Cover image © Andrea Barbieri

Pari Publishing

Via Tozzi 7, 58040 Pari, Grosseto, Italy

www.paripublishing.com

The Pari Dialogues

The Pari Dialogues is the first in an annual series of essays contributed by experts in their fields who are known for the breadth and depth of their knowledge, and who reflect the philosophy of, and the discussions held at, the Pari Center for New Learning. While the essays have been divided into three main areas, they form a unified volume in which each enriches and informs those in other areas.

Visit the Pari Center for New Learning at www.paricenter.com

Contribute to the Pari Dialogues blog at
www.paripublishing.com/blogs/paridialogues

TABLE OF CONTENTS

PREFACE

In 2001 the founders of the Pari Center, Maureen Doolan and David Peat, visited the town of Cordoba in Spain where we learned of its exceptional past. Between the 9th and 13th centuries, Cordoba was one of the key cities of Europe and home to such outstanding thinkers as Averroes, Ibn al-'Arabi, Moses Maimonides and Judah ha-Levi. Under Arabic rule it exhibited great religious tolerance and, thanks to Arab scholars, the writings of Greek philosophers, including Aristotle, reached the West and sowed the seeds for what was to become the Renaissance. Later, under the Spanish king Alfonso X, a university was established where Jews, Christians and Arabs could study side by side. Alfonso also had the Torah and Qur'an translated into Latin so that Christian scholars could study these sacred texts. We also learned of similar communities of scholars in such places as Toledo, and it became clear that the deep collaboration between the three Abrahamic religions arose out of their mutual love of learning and the pursuit of new discoveries in astronomy, medicine, mathematics and botany.

On our return we decided to organize a small roundtable at the Pari Center entitled *A Dialogue between Three Communities and Science* and on the final day of that meeting we applied to the Metanexus Institute to join their Local Societies Initiative program—an initiative which offered funding to invite speakers from the fields of religion and science. (Ironically that last day of our meeting was also the final date for application to Metanexus.)

Our application was successful and over the years that followed we have been able to bring in speakers both from Italy and abroad. As time went by we decided to broaden our scope, not only to explore the connections between religion and science but more widely to include the role of spirituality and the sacred within economics, the arts and society in general. We also organized several conferences (in 2003, *Unlimited Love: Self-Transcendence and Personal-Social Transformation—A participatory learning approach to the future of human well-being*; in 2004, *The Next Horizon: Re-examining Deep Values in Religion & Science* and in 2006, *Ethical Choices in Economics, Society and the Environment*) as well as a number of roundtables.

The Pari Center was represented at the inaugural meeting of SPES—the European Forum for Spirituality in Economic and Social Life. We are also proud to note that we are a three-time winner of the Metanexus Prize.

It seemed entirely appropriate that such topics should be explored in the medieval hilltop village of Pari, Tuscany, for in the medieval period nature, beauty and the sacred were seen as one. For Duns Scotus the universe was "so wondrous as to be compared to a beautiful canticle, a symphony of joy and harmony". William of Auxerre saw the goodness of material things as being one and the same as their beauty and for Dionysius the Aropagite, the cosmos was created out of beauty, and beauty was "the cause of the harmony and splendor in all things, flashing forth upon them all, like light, the beautifying communications of its originating rays". While for Abbot Suger of St-Denis in Paris, the contemplation of colored light, precious stones, fabrics and architectural proportions transported him into a "strange region of the universe", one that lay between earth and heaven. In that same period the artisan, miner, metal worker and artist were considered as the midwives to nature, aiding in her striving to perfection. Indeed the contemporary composer of devotional music, John Tavener, has referred to our "one simple memory", a deeply buried sense of a time when understanding was unified and spoke to our inner wholeness. It is a vision also embraced by the Sufi mystics, as Del Re McWeeny discusses in her essay, where beauty is seen as the manifestation of the Godhead to itself.

For his part, the physicist Wolfgang Pauli looked with deep sympathy to the 17th century when the emerging quantative science of Kepler and Galileo coexisted with a more deeply symbolic approach to nature and matter—each complementing the other. At the same time he feared that, during the 20th century, physics had lost that initial vision to become obsessed with a "will to power" as it sought to exert increasing control over nature. Nevertheless, Pauli felt the "resurrection" of spirit within matter was possible within our own time. The true goal of science, he felt, was not power over nature but, through our enquires, to seek out our own redemption. "Nature expresses itself through wholeness," he wrote, "and the challenge is to experience this wholeness in oneself".

In discussing science and spirituality inevitably the notions of truth and beauty appear. David Bohm often said that the scientist must pursue truth, "no matter where it leads". In many ways this is a religious statement, for the spiritual seeker must always risk the possibility of that "dark night of the soul" and the arid desert of doubt. Likewise, a scientist should have the courage to take that next step which may confirm or destroy his or her life's work. For her part Simone Weil wrote of "that transcendent kingdom to which the only truly great have access and wherein truth abides".

And as to beauty, it is deeply connected to truth, as Keats observed, and clearly enters into the spiritual context, as the quotations given above indicate. Beauty, or elegance, is also a significant factor in mathematics and the sciences. For the eminent pure mathematician, G.H. Hardy, a mathematician is a maker of "patterns of ideas" and should be judged by the criteria of beauty and seriousness in these patterns. For Roger Penrose "one really studies mathematics mainly for its beauty . . . since beauty is an end in itself but it is also a means to that end"; for when we are stuck trying to solve a problem it is best to try to do the most aesthetically attractive thing possible. Likewise P.A.M. Dirac experienced a strong physical sensation when faced with a beautiful result in physics.

There is also the question as to that term itself: "science". Should it be confined only to the sorts of activities associated with the "Western mind" or, as Sean Howard asks, can it be used to embrace other, indigenous knowledge systems?

Issues of authority also surface both in science and in organized religion. These are specifically addressed in George Coyne's essay. Such issues are often linked to a deep desire for security and certainty. But if the science of chaos theory has taught us one thing in the last decades, it is that certainty and control are limited when it comes to the more interesting and complex systems of nature and society. In this way a science that sought to answer everything came face to face with hubris, a lesson that should also be learned at the more rigid levels of organized religion. In the last analysis it is always more exciting and adventurous to remain with the pregnant question than to seek closure with an answer.

It is in this spirit that the *Pari Dialogues* were carried out and it is our intention that each year a new volume of essays be published. The essays presented in this first volume represent some of the talks and discussions that have been held in Pari. In addition, we have also invited some commissioned essays by individuals who were present at our discussions in Pari but chose to write in retrospect.

For this first volume of the Pari Dialogues, we at the Pari Center would like to express our heartfelt thanks to all our contributors and in particular the Metanexus Institute and Eric Weislogel who helped to make this volume happen.

F. David Peat, Pari, 2007

Science and Religion

George V. Coyne, SJ

George Coyne, PhD *is a Jesuit priest, astronomer and Emeritus Director of the Vatican Observatory. He is an active member of the International Astronomical Union, the American Astronomical Society, the Astronomical Society of the Pacific, the American Physical Society and the Optical Society of America. A comet has been named after him.*

ORIGINS AND CREATION

George V. Coyne, SJ

Introduction

It is arguably difficult to find a more heated topic of discussion than that concerning the origins of the universe, and especially of life and of intelligence, and whether such origins can be understood without evoking a Creator God. Responses range from the extremes of a Stephen Hawking or a Pope Pius XII to almost all conceivable intermediate positions. Hawking claims that, if his quantum cosmological theory of the origins of the universe without boundary conditions is correct, then we have no need of God. Pius XII attempted to claim that with Big Bang cosmologies scientists were coming to discover what had already been known from the Book of Genesis, namely that the universe had a beginning in God's creative action. In between we have such positions as evolutionary naturalism and episodic divine intervention. Evolutionary naturalists would claim that, although our scientific knowledge of evolution is limited, the best explanation of the universe and all that it contains is through complexification in an expanding, evolving system in which both deterministic and chance processes play out their roles in a universe abundant with opportunities, 15 billion years old and containing 10^{22} stars (i.e. one followed by twenty-two zeros). Those who profess episodic divine intervention would claim that divine activity is required, at least in some phases of the evolutionary process and, in particular at the occurrence of human life and intelligence, because natural processes alone are not adequate to explain the end result. What is one who is both a believing Christian and a scientist to make of all of this?

Scientific Evidence for Universal Evolution

Let us take a sweeping view of a reasonable scientific picture of things. By reasonable I mean that, while we do not have all the answers, the following picture can stand up to any scientific critique. If we look today in infrared light at the center of Orion we see boiling gas and dust. If we look even closer up we see incandescent regions buried in that gas and with the Hubble Space Telescope we see the fine separation of blue gas and red gas in the midst of a rather chaotic structure. The fact is that stars are being born in this gas. And where the hottest, most massive and, therefore, brightest stars are already born, they are irradiating the gas, and it is giving off hydrogen alpha radiation. In this way we can identify star birth regions. The region of star birth in Orion is just a little part of our Milky Way. Our Milky Way, like most other spiral galaxies, measures 100,000 light years across and it contains about a hundred billion stars. It has several beautiful spiral arms and the sun is located in one of the outer arms, about two thirds of the distance from the nucleus of our galaxy.

How is a star born? It happens by the laws of physics. A cloud of gas and dust, containing about 100 to 1,000 times the mass of our sun, gets shocked by a supernova explosion or something similar and this causes an interplay between the magnetic and gravity field. The cloud begins to break up and chunks of the cloud begin to collapse. And as any gas collapses, it begins to heat up; as it expands, it cools down. In this case the mass is so great that the internal temperature reaches millions of degrees and thus turns on a thermonuclear furnace. A star is born. Thermonuclear energy is the source whereby a star radiates to the universe. You need a very hot piece of the universe to do this, and so you can only get this thermonuclear furnace by having a cloud collapse and raise the temperature. You can only get it, in other words, in stars, with one exception, namely, in the very hot early universe before galaxies or stars were born.

Stars also die. A star at the end of its life can no longer sustain a thermonuclear furnace and so it can no longer resist against gravity. It collapses for a final time, explodes and expels its outer atmosphere to the universe. This may happen nice and peacefully or it may happen in a violent cataclysmic explosion, called a supernova. The most famous of these is the Crab Nebula which has a pulsar at the middle as its dead star.

So stars are born and stars die. And as they die they spew left over star matter out to the universe. The birth and death of stars is very important. If it were not happening, you and I would not be here, and that is a scientific fact. In order to get the chemical elements to make the human body, we had to have three generations of stars. A succeeding generation of stars is born out of the material that is spewed out by a previous generation. But now notice that the second generation of stars is born out of material that was made in a thermonuclear furnace. The star lived by converting hydrogen to helium, helium to carbon, and if it were massive enough, carbon to oxygen, to nitrogen, all the way up to iron. As a star lives, it converts the lighter elements into the heavier elements. That is the way we get carbon and silicon and the other elements to make human hair and toe nails and all of those things. To get the chemistry to make amoebas we had to have the stars regurgitating material to the universe.

Obviously this story of star birth and death is very important for us. Out of this whole process around one star, which we call the sun, a group of planets came to be, among them the little grain of sand we call the Earth. An amazing thing happened with that little grain of sand when, in the 16th and 17th centuries with the birth of modern science, we developed the capacity to put the universe in our heads. We do that by using mathematics and physics, and to some extent the laws of chemistry and biology. Since we have the capacity to put the universe in our heads, a further question comes to us. Where did galaxies come from? All of the material in the universe is concentrated in galaxies and their environs. Galaxies are the building blocks of the universe. Hubble Space Telescope has been able to photograph some of the most distant objects we have ever seen in the universe. They are at a distance of about ten billion light years from us. So we are seeing these objects as they were ten billion years ago. We think that Hubble is seeing proto galaxies. We see, for instance, a case of two blobs which seem to be merging and perhaps building up a galaxy. However, this is very controversial. We are uncertain about galaxy formation, whether it is bottom up with small units that build into a galaxy, or top down with a big cloud that collapses to form a galaxy, and then the stars form within it. Nevertheless, when we compare distant galaxies to nearby galaxies, we see clear differences in the stellar populations. Galaxies as they are born and age go through an evolutionary process. Galaxies are participating in the expansion of the universe. When we look at them on a large scale we see that they are not distributed homogeneously. There are large empty spaces and many dense alignments.

Let us now review what we know of the history of the expanding universe (see Figure 1). As it aged, distances got larger in the universe. As this happened certain key events took place. Quarks combined to form elementary particles, which in turn formed atoms and then molecules. The universe became transparent and the cosmic background radiation came to be. Galaxies and stars were formed. The first microscopic life forms came to be after twelve billion years in a fifteen billion-year-old universe. Why did it take so long to make even an amoeba? We have already discussed one reason. We did not have the chemistry to make even an amoeba until we had had three generations of stars.

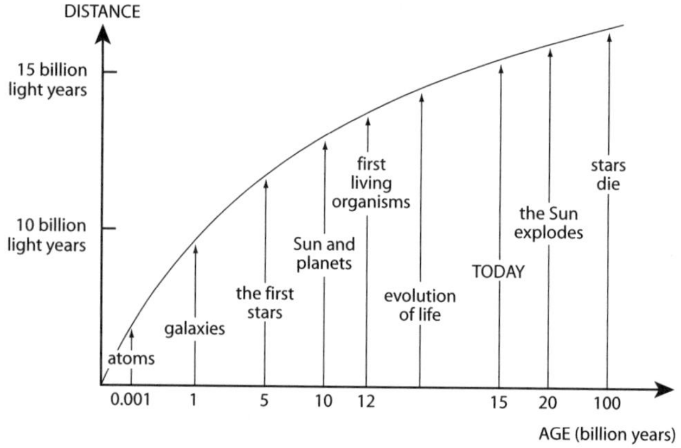

Figure 1: A schematic view of the expanding universe. As the universe grew older (horizontal axis) distances among objects in the universe (vertical axis) grew larger. Key events that occurred are indicated from the formation of atoms to the death of all stars. Note that the first microscopic life forms came to be only after 12 billion years in a universe which is now 15 billion years old.

Human Origins

How did we humans come to be in this evolving universe? It is quite clear that we do not know everything about this process. But it would be scientifically absurd to deny that the human brain is a result of a process of chemical complexification in an evolving universe. After the universe became rich in certain basic chemicals, those chemicals got together in successive steps to make ever more complex molecules. Finally in some extraordinary chemical process the human brain came to be, the most complicated machine that we know.

Let us pause for a moment to review the degree of certainty which we can place in the above scenario. We certainly do not have the scientific knowledge to say how each living creature came to be in detail. We do not know precisely how each more complex chemical system came to contribute to the process of self organization which brought about the diversity of life forms as we know them today. Most importantly, we do not know with scientific accuracy the sufficient elements in nature to have brought about the unbroken geneological continuity in evolution that we propose actually happened. There are, in brief, epistemological gaps which prevent natural science from saying that a detailed theory of biotic evolution has been proven. What we have presented is the most adequate account conceivable at this time considering the available empirical data. And that empirical data, with respect to biotic evolution, comes from various independent scientific enterprises, including molecular biology, paleontology and comparative anatomy.

Did we come about by chance or by necessity in this evolving universe? The first thing to be said is that the problem is not formulated correctly. It is not just a question of chance or necessity because, first of all, it is both. Furthermore, there is a third element here that is very important. It is what I call "opportunity". What this means is that the universe is so prolific in offering the opportunity for the success of both chance and necessary processes that such a character of the universe must be included in the discussion. The universe is 15 billion years old, it contains about 100 billion galaxies each of which contains 100 billion stars of an immense variety. For 15 billion years the universe has been playing at the lottery. What do I mean by the lottery? When we speak about chance we mean that it is very unlikely that a certain event would happen. The "very unlikely" can be calculated in mathematical terms. Such a calculation takes into account how big the universe is, how many stars there are, how many stars would have developed planets, etc. In other words, it is not just guesswork. There is a foundation in fact for making each successive calculation.

A good example of a chance event would be two very simple molecules wandering about in the universe. They happen to meet one another and, when they do, they would love to make a more complex molecule because that is the nature of these molecules. But the temperature and pressure conditions are such that the chemical bonding to make a more complex molecule cannot happen. So they wander off, but they or identical molecules meet billions and billions of times, trillions if you wish, in this universe, and finally they meet and the temperature and pressure conditions are correct. This could happen more easily around certain types of stars than other types of stars, so you can throw in all kinds of other factors.

The point is that from a strictly mathematical analysis of this, called the mathematics of nonlinear dynamics, one can say that as this process goes on and more complex molecules develop, there is more and more direction to this process. As the complexity increases, the future complexity becomes more and more predetermined. There are also necessary, deterministic processes occurring. But there is a lot of chance as to what the exact conditions are when certain molecules meet, so that the necessary process may take place. There are definitely both of these processes, but they are happening in a universe that is so prolific that the eventual outcome has a kind of predetermined nature.

This predetermined nature may be represented by a tree, the Tree of the Universe. It is a strange tree in that everything that ever happened in the universe, from the making of quarks to the making of toenails, is all here. Even those processes that never succeeded, that failed, every dead leaf and dead branch has been conserved. Every meeting of molecules in inopportune circumstances is there. The tree has never been pruned. But blow a quiet breeze through this tree and what will you see? You will inevitably see something that resembles the bare trunk of a tree with certain branching to various forms of life and ending at the top with the human being (see Figure 2). The result is inevitable because with a combination of chance and necessary processes in a very prolific universe with so many opportunities there is a narrowing down of the evolutionary process due to the nature of physics, chemistry, biology and non-linear dynamics.

If we truly accept that there are chance processes involved, then the branching of the tree could be somewhat different. But since complexity proceeds towards an ever more determined direction the trunk of the tree could not be very different. The paleontology, biology and chemistry behind all of this is quite uncertain, but it is clear that something like this would happen. Why is the human being at the top? It is because we are ignorant. We do not know what else to put at the top. The human brain is the most complicated mechanism we know.

Implications for Religious Belief

How are we to interpret this scientific picture in terms of religious belief. Do we need God to explain this? Very succinctly my answer is no. In fact, to need God would be a very denial of God. God is not the response to a need. One gets the impression from certain religious believers that they fondly hope for the durability of certain gaps in our scientific knowledge of evolution, so that they can fill them with God. This is the exact opposite of what human intelligence is all about. We should be seeking for the fullness of God in creation. We should not need God; we should accept him when he comes to us.

The religious believer is tempted by science to make God "explanation." We bring God in to try to explain things that we cannot otherwise explain. "How did the universe begin?", "How did we come to be?" and all such questions. We sort of latch onto God, especially if we do not feel that we have a good and reasonable scientific explanation. He is brought in as the Great God of the Gaps. I have never come to believe in God, nor do I think anyone has come to believe in God, by proving God's existence through anything like a scientific process. God is not found as the conclusion of a rational process like that. I believe in God because God gave himself to me. That was not a miracle. It does make sense that there is a personal God who deals with me and loves me and who has given himself to me. I have never come to love God or God to love me because of any of these reasoning processes. I have come to love God because I have accepted the fact that he first made the move towards me.

The scientific picture traced above deals with the questions of origins, of how what we observe and experience today came to be? The question of creation, and therefore of a God Creator, responds to the question of why is there anything in existence. Creation is not one of the ways whereby things originated as opposed to other ways that can be thought of, including quantum cosmology and evolutionary biology. The claim that all things are created is a religious claim that all that exists depends for its existence on God. It says nothing scientifically of how things came to be, although beautiful stories are told in the Book of Genesis, to elaborate on the dependence of all things for their existence upon God.

Having opened the Pandora's box of the Bible, let us elaborate a bit upon it. The Bible is a collection of writings by various authors at various epochs using various literary genres. And so it best serves reason if one speaks of a specific book rather than of the Bible in general. It is clear that the overall intention of the authors of Genesis is to evoke religious faith, an adherence to the God of Abraham, Isaac and Jacob, and not to teach. There is simply no scientific teaching in Genesis. In the Judaic/Christian tradition, the roots of religious belief reach to 5,000 years before Christ with the prophet Abraham. But Modern Science cannot be dated before the 16th or 17th century, roughly from the time of Galileo and then through many others to Newton, with the discovery of the universal law of gravity, the differential calculus, etc. You may even wish to go back to the beginnings of the experimental method with Roger Bacon and others in the 13th century. But, at any rate, the modern science that speaks to religion today is born much later than the religion to which it speaks. It has to be recognized that the religious tradition is historically much longer and to a certain extent has that richness of the past that modern science does not.

It is unfortunate that, at least in America, creationism has come to mean some fundamentalistic, literal, scientific interpretation of Genesis. Judaic-Christian faith is radically creationist, but in a totally different sense. It is rooted in a belief that everything depends upon God, or better, all is a gift from God. The universe is not God and it cannot exist independently of God. Neither pantheism nor naturalism is true.

But if we confront what we know of origins scientifically with religious faith in God the Creator, in the senses described above, what results? I would claim that the detailed scientific understanding of origins has no

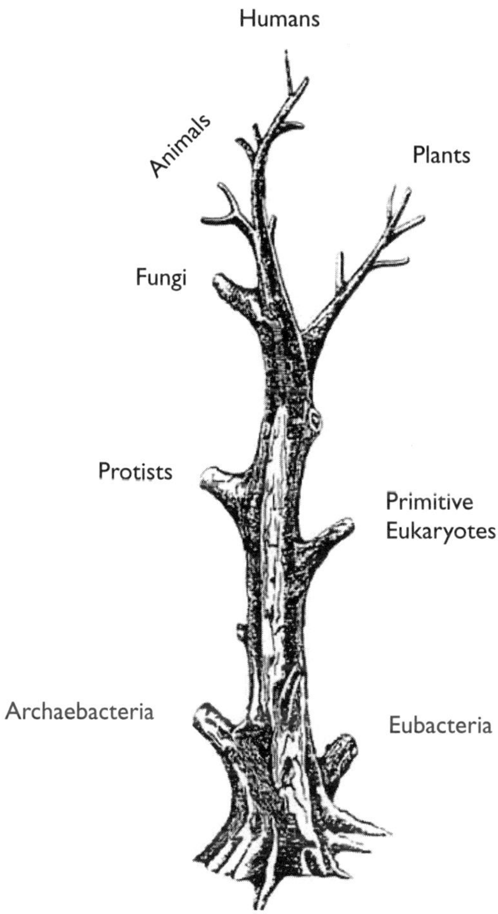

Figure 2. If we represent the evolution of the universe as a tree and we take only that part of the tree which led to human life and clean it of all the processes which did not succeed, we will inevitably see something like this. It could have been somewhat different, because there were chance processes involved. But it could not be very different because there were also necessary processes and chemical complexification channels the evolutionary process towards the human brain and, indeed, beyond.

bearing whatsoever on whether God exists or not. It has a great deal to do with my knowledge of God, should I happen to believe he exists. Let me explain.

Take two rather extreme scientific views of origins: that of Stephen Jay Gould of an episodic, totally contingent and, therefore, non-repeatable evolutionary process as contrasted to a convergent evolutionary process such as that of Christian de Duve, in which the interplay of chance, necessity and opportunity leads inevitably to life and intelligence. In either case, it is scientifically tenable to maintain an autonomy and self-sufficiency of the natural processes in a natural world, so that recourse to God to explain the origins of all that exists, is not required. It is not a question of chance in nature, excludes God; destiny in nature requires God. In neither case is God required.

If, however, I believe in God then what nature tells me about God in one case is very different from what nature tells me about God in the other. Please note that I am not calling upon faith to adjudicate between contrasting scientific viewpoints. I do think that convergent evolution is more consistent with God's revelation of himself in the Book of Scripture, so that, as Galileo was fond of stating, the Book of Scripture and the Book of Nature speak of the same God.

If we take the results of modern science seriously, it is difficult to believe that God is omnipotent and omniscient in the sense of the scholastic philosophers. Science tells us of a God who must be very different from God as seen by the medieval philosophers and theologians. Let us ask the hard question. Could, for instance, God after a billion years in a fifteen billion-year-old universe have predicted that human life would come to be? Let us suppose that God possessed the theory of everything, knew all the laws of physics, all the fundamental forces. Even then could God know with certainty that human life would come to be? If we truly accept the scientific view that, in addition to necessary processes and the immense opportunities offered by the universe, there are also chance processes, then it would appear that not even God could know the outcome with certainty. God cannot know what is not knowable.

This is not to place a limitation upon God. Far from it. It reveals a God who made a universe that has within it a certain dynamism and thus participates in the very creativity of God. Such a view of creation can be found in early Christian writings, especially in those of St. Augustine in his comments on Genesis. If they respect the results of modern science, religious believers must move away from the notion of a dictator God, a Newtonian God who made the universe as a watch that ticks along regularly. Perhaps God should be seen more as a parent or as one who speaks encouraging and sustaining words. Scripture is very rich in these thoughts. It presents, indeed anthropomorphically, a God who gets angry, who disciplines, a God who nurtures the universe. Theologians already possess the concept of God's continuous creation. I think to explore modern science with this notion of continuous creation would be a very enriching experience for theologians and religious believers. God is working with the universe. The universe has a certain vitality of its own like a child does. It has the ability to respond to words of endearment and encouragement. You discipline a child but you try to preserve and enrich the individual character of the child and its own passion for life. A parent must allow the child to grow into adulthood, to come to make its own choices, to go on its own way in life. Words which give life are richer than mere commands or information. In such wise does God deal with the universe.

These are very weak images, but how else do we talk about God? We can only come to know God by analogy. The universe as we know it today through science is one way to derive analogical knowledge of God. For those who believe modern science does say something to us about God. It provides a challenge, an enriching challenge, to traditional beliefs about God. God in his infinite freedom continuously creates a world which reflects that freedom at all levels of the evolutionary process to greater and greater complexity. God lets the world be what it will be in its continuous evolution. He does not intervene, but rather allows, participates, loves. Is such thinking adequate to preserve the special character attributed by religious thought to the emergence of spirit, while avoiding a crude creationism? Only a protracted dialogue will tell.

Virginia Del Re McWeeny

Virginia Del Re McWeeny *read Sanskrit, art, religions and literature of Ancient India and the East; classical literature; and modern languages, French, English and Farsi at the University of Rome and traveled extensively in Tehran, Iran, Lebanon, Turkey, Ghana, Nigeria, Togo and Dahomey. Throughout her life she has pursued her interest in Indian and Persian studies and has been translating the Persian Mystic Poets. Del Re McWeeny is also an active member of Associazione Casa della Donna di Pisa (an association of women) and works on women's social issues, and women's literature.*

THE WONDER OF CREATION

A few notes for reflection on Sufism and science and some "correspondences"

VIRGINIA DEL RE MCWEENY

> The day will come when, after harnessing space, the winds, the tides, gravitation, we shall harness for God the energies of love. And, on that day, for the second time in the history of the world, man will have discovered fire.
>
> Among so many great events, there is one phenomenon which, in the eyes of posterity, may well overshadow everything that has been discovered in radiation and electricity: the permanent entry into operation, in our day, of inter-human affinities—the movement, irresistible and ever increasing in speed, which we can see for ourselves, welding peoples and individuals to one another . . .
>
> Pierre Teilhard de Chardin[1]

While attending some of the Dialogues between Science and Religion in Pari, my mind went to the great mystical (or 'Sufi') Persian poets of the Middle Ages I had been recently writing about[2] (from an entirely literary angle) and to their position *vis-à-vis* science. Also some ongoing scientific and political developments have made me think of the many different roads travelled by human thought, of the "other"—as compared to science and "reason"—ways of experiencing the material and the immaterial world, different approaches that keep reappearing, sometimes in startlingly unexpected forms, building unimagined bridges between apparently irreconcilable worldviews and modes of knowing, and between past and present.

I asked myself: what is—or was—the relationship between Sufism and science? Is coexistence possible or even some sort of conciliation? Can the paths of science and mysticism cross in the modern world? Can we even pose such a question? Because, of course, to speak in the same breath of mysticism and science sounds, at first, altogether absurd.

Two main considerations persuaded me that the question about Sufism and science is not futile. One was the fact that Sufism, or Islamic mysticism, is a widespread spiritual movement in the modern world. Although it rests on a longstanding tradition of great masters, poets and philosophers, which reached its peak in the period between the 10th and the 16th century, there are traditional and modern Sufi groups and Orders, with many thousands, sometimes millions of adepts.[3] One of the best known and popular among the Western public is the modern syncretistic kind of Sufism, especially the International Sufi Movement and the International Sufi Order whose adepts draw inspiration from the teachings of their Master, Hazrat Inayat Khan (Pir-o-Murshid Inayat Khan, 1882-1927),[4] in an effort of ideal synthesis between East and West. Thus, Sufism is still an enduring, significant power, not only within the Islamic religious context, from which it arose and to which it firmly belongs, but as a source of inspiration for those people who strive to transcend the particular doctrines and dogmas of positive, organized religions, accepting only those common essential elements in them whereby souls share the same spiritual pursuit: what Teilhard de Chardin refers to, poetically, as the *inter-human affinities*. And certainly their endeavour responds to a particularly pressing need in a world where differences and destructive conflict threaten to defeat all endeavours towards peace.

This discrete, yet extremely influential presence seemed to me one good reason that justified the question about Sufism and modern science and the ethical and philosophical problems it poses.

The second, perhaps more important, reason that can lend legitimacy and relevance to the question, even for non scientists like me, is that science today, or, more simply, many scientists, no longer eschew issues that are frankly metaphysical, and appear, in fact, anxious to find ways of reconciling scientific pursuits and faith in a transcendent reality. We can all see how the need for meaning becomes more and more desperate as the frontiers of knowledge move farther and farther and seem to fade away altogether, and the limitations of human reason appears in all its anguishing evidence.

Sufism and science, then: let's say first of all that Sufism, classic or modern, does not reject the world's physicality and the reality of sensorial objects; nor does it deny or despise the pure sciences, nor scientific discoveries and technological advancement, or philosophical and theological speculation. It does not refuse or even dispute all that, but stands outside it, transcending it in the name of a *quality*, a different *essence* of human knowing and experiencing, where the object of enquiry and the informing/normative principle coincide, eluding absolutely the "investigation of reason". Essentially, Sufism pursues *sapientia* rather than *scientia*. The mystic's truth and the scientist's truth, in all places and at all times, diverge profoundly in their object, in their aims and in their methods. The mystic's quest is for the direct apprehension of Truth, i.e., the Godhead, while scientists pursue the truth that can be attained through their intellect and their tools; the mystic's methods are the refining of the heart, those of the scientist the refining of tools and techniques of intellectual investigation. The only *laws* in the mystic's field of enquiry are the heart's desire, the renunciation of the selfish Ego, and above all, Love. Sufism is in fact also described as "the heart's science" or "the path to God through love". As Rumi, the greatest mystic poet of Islam, said:

> Though it be as precious as pearl or coral, reason's enquiry
> is all different from the soul's search! [...]
> The enquiry of reason and of the senses—let it be known—is about causes or effects:
> the soul's enquiry is through wonder or miracle.
> The Soul's light came: it did not stay: O you who want to be enlightened,
> [do not look for] either "consequence", or "premiss", or "implication" or "exclusion".
> Because the seer whose light now resplends, has no need for arguments,
> nor for somebody to lead by a stick [like a blind man].[5]

Rumi's lines explain perfectly and synthetically the attitude of the mystic towards "rational" science. (But let's note, in passing, how the terms he uses reveal an excellent knowledge of the technical terms of Logic.)

The Sufi path does not coincide with the path of science: and this is true of all mysticism. The objects of their search could not be farther apart: the scientist moves in the physical world to observe it and investigate its laws, and to put them, where possible, to the use of humankind. The mystic lives in the world, and yet moves towards a hidden reality, a sphere to which he knows he belongs and to which he longs to return. Finding that truth, however, implies entering a dimension where human space and time are utterly meaningless. Science, as we see it today, demands proofs and verification, demonstration under well established rules and procedures and, while it does not admit limits to its possibilities of investigation in the physical world, is very careful to keep away from "absolute truths". Mystics—as well as religious believers, and even lovers and poets—speak of things that by definition escape human reason and its means, and of which no scientific demonstration is possible. And yet, today's physics is again posing questions about the Ultimate Causes, as does biology, whose frontiers seem to be recklessly getting closer and closer to the origin of life, to creation . . .

Not being a scientist myself, I won't even marginally touch upon scientific topics, nor shall I attempt comparisons or links not only arduous in themselves but, most probably, historically unjustifiable. Having said that, two or three points in two great poets of the classic era of Sufism, Rumi and Jami, and in some Sufi philosophers such as Shurawardi—the Master of Illuminativism—seem to be extraordinarily topical and relevant in the context of a possible discussion about Sufism and modern science, as they find striking "correspondences" with ideas expressed by some modern scientists.

I will present them here just as possible points for reflection, like sparks from a past that suddenly seems very close.

First a brief digression: a whole page of the cultural Sunday supplement to the Italian newspaper *Sole 24 Ore* (a very close equivalent to the British newspaper *Financial Times*), dated January 16, 2005, carried—in block letters—a question: "what do you believe is true even though you cannot prove it?" This was the year's "edge" question John Brockman had just put on line (www.edge.org) to a great number of intellectuals. The Italian paper published nine out of the 120 answers Brockman got: one by a writer, two by physicists, two astrophysicists, two biologists and two psychologists. Each one of them admitted to a belief—that is, an absolute certainty—in something he cannot offer scientifically valid proofs or demonstrations for, even within his own field. Some were writing in a slightly embarrassed or semi-playful tone, but none of them was shocked by the question, nor denied its legitimacy. Perhaps it is true that we all have strong certainties about non-demonstrable, even irrational things, and also that to deny this fact is actually the most irrational of all our beliefs.

To my surprise, two of the answers made me think immediately of the Sufi: Leon Lederman, Physicist and Nobel Laureate, said he believes in the essential beauty of Nature, though he cannot prove it scientifically. I quote from his answer:

> To believe without knowing it cannot be proved (yet) is the essence of physics. Guys like Einstein, Dirac, Poincaré, etc. extolled the beauty of concepts, in a bizarre sense, placing truth at a lower level of importance.

Lederman goes on to tell of the

> . . . pain [in the scientific community] at the loss of simplicity and harmony [when] the long-respected law of mirror symmetry was violated by weakly interacting but exotic particles: [a pain] alleviated by the discovery of the failure of particle-antiparticle symmetry [which seemed to restore] a new and more powerful symmetry. . . . How silly of us . . . to have lost confidence in the essential beauty of nature! . . . Surely, there is in store some spectacular, new, unforeseen splendor in all of us. She [the Creator] will not let us down. This we believe, even though we can't prove it.

David Buss, a psychologist, believes—against all odds—firmly in love. This is what he says:

> I've spent two decades of my professional life studying human mating. In that time, I've documented phenomena ranging from what men and women desire in a mate to the most diabolical forms of sexual treachery. [...] But throughout this exploration of the dark dimensions of human mating, I've remained unwavering in my belief in true love.
>
> While love is common, true love is rare, and I believe that few people are fortunate enough to experience it. The roads of regular love are well traveled and their markers are well understood by many—the mesmerizing attraction, the ideational obsession, the sexual afterglow, profound self-sacrifice, and the desire to combine DNA. But true love takes its own course through uncharted territory. It knows no fences, has no barriers or boundaries. It's difficult to define, eludes modern

measurement, and seems scientifically wooly. But I know true love exists. I just can't prove it.

As Ahmad al-Ghazali, (see below and ref. 6) said a long time ago, "... the distinctive orientations of love are incidental. By its essence, it transcends all directions because it needs not turn into any direction in order to be love."

Baraka

Now, Beauty and Love are two fundamental tenets of Sufism that find justification and support in Muhammad's saying, "God is beautiful and loves Beauty" and are inextricably connected to one another. Rumi endlessly sings Love and Beauty in Nature and in the human being, as do Attar and Jami in their verses. To contemplate Beauty is to love, and vice versa: there lies the way to Knowledge and to the union with the Godhead. The "faithful lovers" (some will remember the *fedeli d'amore* of medieval times) render their divine service to Beauty by contemplating it as the greatest of all theophanies. Ahmad Ghazali (the great Persian mystical preacher, brother of Abud Hamid Ghazali, the theologian who is somehow considered responsible for the end of scientific research in Islam) writes a celebrated treatise on Love, *Sawanih*.[6] But being a mystic does not mean, then as now, being deaf to "rational" knowledge and to the physical nature of the world. Indeed the great mystics of the so-called golden age of Islam were often learned men, well versed, not only in theological and philosophical matters, but in mathematics, astronomy and the natural sciences, such as medicine, botany, and geography (the *Philosophia Naturalis* sciences, all at the time classified in hierarchical order under theology). Hamadani, Ahmad Ghazali's favourite disciple was a mystic and a judge, a mathematician and a philosopher.

But the Sufis, rather than following the devouring need to describe the world and its phenomena and explain their causes through intellectual means, choose the other road open to those who are anxious to "understand" the Universe: they merge themselves in its indescribable beauty until they are totally penetrated by the sense of the divine presence, the emotion of *Baraka*, that mixture of wonder, adoration and awe, which comes from the contemplation of the unutterable harmony of the Whole and the sign of the One.

From the intellectual standpoint, what today is known as the "intelligent design" behind the phenomenal world is for the Sufis an undeniable and sufficient *datum*: the creation is the work of God, the One, from whom everything derives: indeed there is often in Sufism an attitude, a sensibility that could easily be described as pantheistic: not only the Whole comes from the One, but the One *is* the Whole. This attitude did actually expose them more than once to accusations and to capital sentences for blasphemy and heresy.

However, leaving alone religious belief and theological thought, there are some interesting points in the "philosophers of Illumination" (Najm Razi, Suhrawardi, Najm Kobra, and Mohammad Lahiji, whose writings are somehow a prelude to the work of the great theologian and master of Islamic gnosis, Ibn al-'Arabi) and some ideas about the Creation, Nature, Man and Evolution in Rumi, which appear at least, shall we say, intriguing, in the light of modern science.

Beauty and Light

For many Sufi mystics, saturated with the metaphysics of great philosophers, such as Ibn al-'Arabi, the created world is brought into existence by the desire of uncreated Beauty to express itself, to be known. Jami sings:

Such urge arose first from Eternal Beauty.
She moved Her tent outside the sacred bounds
Revealed Herself to souls and to horizons.
A different face appeared in every mirror.
Her Being was discussed in every place.

Jami, *Yusuf va Zuleika* (transl. J.T.P. De Bruijn)[7]

Hidden behind the veil of mystery, Beauty is eternally free from
the slightest stain of imperfection. From the atoms of the world,
He created a multitude of mirrors; into each one of them He cast
the image of His Face; to the awakened eye, anything that appears
beautiful is only a reflection of that Face.

Jami (transl. A. Harvey and E. Hanut)[8]

And, again:

> From all eternity the Beloved unveiled His beauty in the solitude of the Unseen;
> He held up the mirror to His own face, He displayed His loveliness to Himself.
> He was both the spectator and the spectacle: no eye but His had surveyed the universe.
> All was One, there was no duality, no pretence of 'mine' or 'thine'.
> The vast orb of Heaven, with its myriad incomings and outgoings, was concealed in a single point.
> The Creation lay cradled in the sleep of non-existence, like a child e're it has breathed.
> The eye of the Beloved, seeing what was not, regarded non existence as existent.
> Though He beheld His attributes and qualities as a perfect whole in His own essence,
> Yet he desired that they should be displayed to Him in another mirror,
> And that each of His eternal attributes should become manifest accordingly in a diverse form.
> Therefore He created the verdant fields of Time and Space and the life-giving garden of the world,
> That every bough and fruit might show forth His various perfections.
> The cypress gave a hint of His comely stature, the rose gave tidings of His beauteous countenance.
> Wherever Beauty peeped out, Love appeared beside it; wherever Beauty shone in a rosy cheek, Love lit his torch from that flame.
> Wherever Beauty dwelt in dark tresses, Love came and found a heart entangled in their coils.
> Beauty and Love are as body and soul; Beauty is the mine and Love the precious stone.
> They have always been together from the first: never have they travelled but in each other's company.

Jami (transl. R. A. Nicholson)[9]

Jami's words can obviously be appreciated simply as beautiful poetry on its own merit; but, as we have seen, their meaning rests on a very profound philosophy of Creation whose roots can be found in Islamic theosophy and, further back, in Hermetism, Neoplatonism and in the old Iranic religions. Take for instance the recurrent symbolism of the mirror: on the one hand, it is connected to the typical Sufi concept of the heart as the mirror of true reality, i.e., the *inner* reality, as opposed to the *external* reality, that is pursued and perceived by the senses and by the *discursive* intellect. The human intellect, that setting on tortuous and treacherous tracks, loses sight of the Absolute Truth, the One, and thus becomes guilty of arrogance, of *hubris*. As David Peat[10] reminds us, today's scientists, when formulating a theory or making an observation, do not take into account their inner being, nor their own ethical needs.

On the other hand, the mirror is naturally connected to the idea of light, as the instrument that reflects and refracts Light; and the One *is* Light. Allah is "The Light of the Heavens and of Earth", and is "Light upon Light" (Qur'an, XXIV, 35). Light and all that is related to it is an essential symbolic key in the Sufi mystical lexicon (as well as in mysticism in general). Light is warmth, love, fire, consummation in passion: the moth—in a frequent metaphor in the mystics' language—cannot stay away from the candle and dies burning in its flame. Christian mystics speak of the "ocean of light" and of the "blinding light in the moment, *unio mystica*".

The modes of this Light that contemplates itself and that, entirely freely, i.e., "as an act of Love", creates the World and in the end its masterpiece, the Human Being, are part of a complicated "metaphysics of light and colours" whose exponents are Suhrawardi, Najm Razi, Najm Kobra and Mohammad Lahiji. Light also enters in the fundamental Islamic idea of the Perfect Man, *The Man of Light*, in this context.

Clearly, the metaphysics of Light is inscribed in the sphere of the *via mystica ad deum*, but there are aspects of it that seem to echo modern physics concepts. On the surprising presence of the Black Light and of light without matter I shall quote briefly the great scholar Henry Corbin, who has written profoundly on this subject.[11]

> The dimension of superconsciousness—as Corbin calls it—is symbolically heralded by the 'black light'; according to Najm Razi and Mohammed Lahiji, this constitutes the highest spiritual stage; according to Semnani, it marks the most perilous initiatic step, the stage immediately preceding the ultimate theophany, which is heralded by the green light. [...] The idea of the 'black light' (Persian *nur-e siyah*) is above all what obliges us to distinguish between two dimensions which could not be accounted for by a one-dimensional or undifferentiable unconscious. To the extent that the mystical language comes to 'symbolize with' physical experience, it seems that the latter perfectly illustrates the idea of a

> polarity not so much between consciousness and the unconscious as between a superconsciousness and a subconsciousness.
>
> There is one darkness which is matter,—Corbin continues—and there is another darkness which is an absence of matter. Physicists distinguish between the *blackness* of matter and the *blackness* of the stratosphere.
>
> On the one hand there is the *black body*, a body that absorbs all light without distinction of color; this is what is "seen" in a dark furnace. When heated, it passes from black to red, then to white, then to white-red. All this light is light absorbed by matter and re-emitted by it. This is also so in the case of the "particle of light" (the man of light*)* absorbed in the dark well which according to Najm Kobra and Sohravardi, is compelled by the fire of the *dhikr* to liberate the particle, to "re-emit" it. This then is the black figure, the well or dark furnace; it is the lower darkness, the infraconscious or subconscious.
>
> But there is another light, a *light-without-matter*, which becomes visible when released from this already made matter that had absorbed it. The darkness above is the blackness of the stratosphere, of stellar space, of the black Sky. In mystical terms, it corresponds to the light of the divine Self in-itself (*nilr-e dhat*) . . .

In 1997 William C. Gough[12], published an interesting article on the possible scientific value of some fundamental Sufi concepts, among which are those precisely related to Light. But, as far as I know, his article does not refer to this "light without matter, nor to the Black Light".

The theosophy of Illumination of Shihāb al-Din al-Suhrawardī is a complex and fascinating system: in the words of Annemarie Schimmel:[13] "for Suhrawardi 'existence is [equal to] light'. This absolute light reaches the created world through innumerable vertical and horizontal orders of angelic entities. The archetype of humanity among the angels is Gabriel, and all things are brought to life by the sound of Gabriel's wings. It is Man's task to recognize the existential light and to come close to it: the more human beings get free from the obscurity of their Ego and let themselves be permeated by light, the closer they get to the Godhead."

The reference to "the beating of Gabriel's wings", so poetically suggestive in itself, is obviously rich with neoplatonic and gnostic reminiscences and goes back even to Zoroastrian angelology, as well as to the angels of Judaism and Christianity. But it can also remind us of a biologist of our time, Rupert Sheldrake, who actually caused a sensation when he spoke of the possible existence of angels, not metaphorical, but real, as superior intelligences operating in our universe. Right or wrong in the scientific sense, Sheldrake is a clear example of the attempts at conciliation that many modern scientists are making between rationalism and spirituality. They are people for whom the unexplainable cannot be just something "not yet explained or discovered".

On this point I'll quote Sheldrake again later, but let's first go back to angels for the moment: Hal Blacker,[14] who interviewed the scientist in 1996, tells how he "met pioneering biologist Rupert Sheldrake the night he and theologian Matthew Fox celebrated the publication of their new collection of dialogues, *The Physics of Angels*. I knew that Sheldrake was not afraid to challenge orthodoxy by entering realms of thought usually eschewed by other scientists. A former Research Fellow of the Royal Society and former Director of Studies in biochemistry and cell biology at Clare College, Cambridge University, his most unorthodox work is not easily dismissed, even by his more traditional peers." "Rupert Sheldrake" writes Blacker, "showed a quality that is rare in men of his intelligence and breadth of knowledge—a pervasive humility and respect for what is not known and for that which it may never be possible for the intellect to grasp."

Here is Sheldrake speaking in Blacker's interview:

> Right from the beginning, since my book *A New Science of Life* was published, my aim has been to try to find a wider picture or paradigm for science that is not constricted to an inanimate, mechanistic view of things.
>
> . . . One approach to this "bigger picture" is the idea of the whole universe as a living organism. . . . Then the next question that arises is: Well, if the universe is alive, if solar systems are alive, if galaxies are alive, if planets are alive, are they conscious?
>
> And is the kind of life that may exist in the cosmos more conscious than ourselves, or do we have to assume it's a great deal less conscious than ourselves? Are we the smartest beings in the universe? Now the usual answer of science is yes. I think that's a very improbable assumption. So, if we come to the idea of many forms of consciousness, if the galaxy has a life and a consciousness, then it would be a consciousness far greater than our own—greater in extent, greater in its implications and power, and greater in the spread of its activities. This, from the point of view of science, is a ridiculous idea, because science has wiped out consciousness

> from everything in the universe except human brains.
>
> But in the Christian tradition, in the Jewish tradition, and in all traditions, there is the idea of many beings with greater levels of consciousness than our own. In the Western traditions, they are called angels. So, in my book with Matthew Fox, *The Physics of Angels*, our aim was to explore what the Western tradition actually has to say to us about angels, and see what relevance that might have in the context of the new cosmology.
>
> My interest is in a new view of science in which we see the universe as alive, and in an exploration of what it could mean to see that there are forms of consciousness above the human consciousness. If one thinks of a divine consciousness as embracing all things, and then this human consciousness here, the traditional view is that there are many, many other levels and kinds of consciousness in between. It's not that you leap straight from divine consciousness to human consciousness, with nothing but brute matter in between.[15]

But, to go back to Sufism: how can the scientific method, the inviolable principle of demonstrability and verification of all hypotheses, and the permissibility, for the sake of knowledge, of all scientific exploration, be reconciled with the world of the *undemonstrable* par excellence, with the direct knowledge of the Godhead, with the *via mystica?*

I have already quoted Rumi's lines on "reason's enquiry", but here I would like to focus on the words "wonder", "thaumaturgy"— the reverential emotion in the face of something miraculous, take us back to the word *baraka* and do not in any way contrast with the feelings expressed by the scientists in their creative work. The perception, indeed the need for harmony and beauty in their pursuits have been and still are constant factors for many of them, for Einstein, for instance, just to mention the most famous among them.[16]

Rumi speaks of the "two kinds of human intelligence":

> There are two kinds of intelligence: One acquired,
> as a child in school memorizes facts and concepts
> from books and from what the teacher says,
> collecting information from the traditional sciences
> as well as from the new sciences.
>
> With such intelligence you rise in the world.
> You get ranked ahead or behind others

in regard to your competence in retaining information.
You stroll with this intelligence
in and out of fields of knowledge, getting always more
marks on your preserving tablets.

There is another kind of tablet, one
already completed and preserved inside you. A spring
overflowing its springbox. A freshness in
the center of the chest. This other intelligence
does not turn yellow or stagnate. It's fluid, and it doesn't
move from outside to inside through the conduits of
plumbing-learning.

This second knowing is a fountainhead from within you, moving out.

Rumi, Mathnavi[17]

Since analysis and logic have proved themselves to be such efficient tools in the advance of scientific knowledge,—writes F.C. Happold[18]—there has been a tendency among the heirs of Western European culture more and more to regard them as the only valid tools of knowledge. They are, however, not the only tools of knowledge.

Our medieval ancestors acknowledged two valid ways of knowledge. The first was through *ratio,* i.e. discursive reason, the second through *intellectus.* The word *"intellectus"* is not easy to define exactly. It carried something of the meaning of intuition or creative insight or imagination, in the sense Blake used that word. *Intellectus* they considered a higher faculty of the mind than *ratio;* one capable of bringing men to a more profound knowledge than could be gained through discursive reason.

The status of intuition, creative insight, imagination, call it what you will, is an acknowledged one. It is recognized not only by the mystic, the poet, and many philosophers, but also by the scientist. It was one of the eminent of quantum physicists, Max Planck who wrote in his autobiography: "When the pioneer in science sends forth the groping fingers of his thoughts, he must have a vivid, intuitive imagination, for new ideas are not generated by deduction, but by an artistically creative imagination".

Rupert Sheldrake, on his part, says, at the end of his conversation with Blacker (see above):

> It's not a particular set of ideas or doctrines that constitutes science. It's a method of inquiry, the idea of building on what has gone before and exploring by experiment, and also openness to new ideas. And that, I think, is completely compatible with a spiritual view of things. I don't think you can prove some of these spiritual truths by scientific means. Science is a limited method of inquiry. It looks at the repetitive aspects of the natural world, so its sphere of interest is relatively confined. Spiritual experience would involve the limits of consciousness and the nature of consciousness. So spiritual inquiry has a broader sphere and science a narrower sphere. But I don't see any incompatibility between the two.

Can then the two ways co-exist; has the conflict between the two modes of knowing no reason to be; can science and faith in the unprovable accept and respect each other's sphere, the difference without the contrast?

To quote Happold again:

> Unlike the materialists of the eighteenth and nineteenth centuries, twentieth-century scientists no longer regard their domain as all-inclusive, embracing everything in heaven and earth: They recognize that it is confined within relatively narrow limits. The results of recent specialist scientific activities have compelled them to conclude that it is necessary to consider orders of reality which are not amenable to the techniques and vocabulary of science. The scientist now [...] does not consider it inconceivable, indeed it seems to fit in with his experience, both as a man and a scientist, that in those regions of personality which lie outside the orbit of normal consciousness the categories of time and space may be inappropriate and that man in his wholeness dwells in a realm that comprehends infinity. The world of his normal experience may be illusion, not in the sense that it does not exist or is necessarily mere appearance, but in that what he perceives is only a section of a greater whole and, wholly or in part, the mental construction of a limited group of perceptions.[19]

Happold was writing in 1970. I do not know if his words about the composition of the conflict between scientific truth and the "other" truth, between the two "methods" of knowledge are too optimistic at the level of our everyday, concrete experience. The choices that the new boundaries of science and technology pose to all of us require much more than a respectful distance between the two areas of "enquiry", and the risk of a global head-on ideological clash seems to be steadily increasing.

I come now to the last point in Rumi's thought that sounds rather extraordinary in the light of modern science. Once again, we are in the heart of Rumi's "oceanic" mystic encyclopedia, his Mathnavi:

> I died as mineral and became a plant,
> I died as plant and rose to animal,
> I died as animal and I was Man.
> Why should I fear? When was I less by
> dying?
> Yet once more I shall die as Man, to soar
> With angels blest; but even from angelhood I must
> pass on: all except God doth perish. When I have
> sacrificed my angel-soul,
> I shall become what no mind e'er conceived. Oh, let me not exist! for
> Non-existence
> Proclaims in organ tones. "To him we shall return."[20]

Elsewhere Rumi explains this (mystical) theory of evolution:

> First Man entered the world of inorganic things. From that stage he passed to the vegetable kingdom where he spent many years, remembering nothing of his previous status, so radically opposed to the present one. And then, as he went on from the vegetable world to the animal world, the vegetable kingdom where he had lived was utterly wiped out of his memory, except for his mysterious attraction towards plants, especially in the season of Spring and of gentle flowers, like a child attracted to his mother. And then the Creator, whom you well know, guided him from the animal to the human state. Thus he moved from a boundless country to another, until he became intelligent and wise and powerful, remembering nothing, now, of his old souls. But he still must migrate from this human intelligence, away from this intellect made of greed and selfishness, he must flee and contemplate a hundred thousand intelligences more beautiful and sublime.[21]

Now, with all due difference, of course, science has for a long time accepted the idea that neither human beings, nor other animals, nor plants sprung suddenly into existence; the Chimpanzee genomic map, we are told, shows a striking similarity to that of Man. Some feel outraged and diminished by this fact, as the old arrogance of the "lord of the Earth" refuses to die . . .

In fact, the true dividing line between those who believe in God and the atheists is whether there is a supreme being who designs and orders, however remotely and indirectly, the universe, including human beings, or if everything is just the result of blind Chance.

The idea of the "First Motor", the universal and eternal Mind or Spirit from whose creative impulse all the material and immaterial world came into existence in its variety of forms and phenomena was familiar to medieval philosophers and even now does not—again, with all due differences—contrast with the conscience of scientists who have a religious belief. And some biologists see evolution in a way that perhaps the ancient mystics—and even less present-day mystics—might not reject. I am thinking of Teilhard de Chardin, a scientist (biologist and anthropologist) and a Christian mystic (but the Catholic Church didn't like his theories at all). He said, "To ask if the universe is still developing means to decide whether the human spirit is or is not undergoing evolution. Now, to this question, I would answer, without any hesitation: yes."

On Teilhard de Chardin's thesis, F.C. Happold[22] writes:

> As seen within the time process, evolution takes the form of a series of unfoldings: from the building up of the universe, through the formation of the earth, to the appearance of life on this planet. Through countless ages the evolution of living forms has gone on, until the next stage, that of the emergence of Man, was reached, first the pre-hominids and finally *homo sapiens.*
>
> With the appearance of the human line a new form of biological existence emerges; there is, in effect, a new creation. This new creation is characterized by the emergence of the power of reflection. Father Pierre Teilhard de Chardin, in his *Phenomenon of Man,* coins the vivid word *hominisation* to describe this leap in the evolution of life. With the appearance of man, instinct is more and more supplemented by the greater and greater interiorization; the *within* more and more predominant. Invention more and more replaces chance as the controlling evolutionary direction of attraction and repulsion supplants that primitive type which operated at the levels of pre-life and the lower forms of life. [...]
>
> The process of evolution may thus be envisaged as the adding of successive "layers", or "envelopes" to the globe. With the appearance of *homo sapiens* to the *biosphere,* i.e. of life, is added the *noosphere,* the envelope of thought, another "membrane to the majestic assembly of telluric layers".

> If this picture of the course of evolution be a true one, what line may evolution be expected to follow? May it not be the growth of an ever higher form of consciousness, spreading out ever wider and wider, until it embraces more of mankind, a greater and greater intensification of *noogenesis,* an expanding interiorization and spiritualization of man, which will result in an ability to see aspects of the universe as yet only faintly glimpsed? And, if that be so, may we not see in the mystics the forerunners of a type of consciousness which will become more and more common as mankind ascends higher and higher up the ladder of evolution? It is not an irrational hypothesis.[23]

Not surprisingly, Teilhard's hypotheses met violent opposition not only from the Church, but from most of the scientific community as well. And yet, they have been recently taken up again by several scientists; in Italy, for instance, by L. Galleni, the Professor of General Zoology at the University of Pisa.

And here again is Rupert Sheldrake on the same theme of creation and evolution:

> The old idea of the earth as dead has given way to Gaia, the idea of the living earth. The old idea of the universe as uncreative has given way to the idea of creative evolution; first in the realm of living things, through Darwin, and now we see that the whole cosmos is in creative evolution. [...] Creativity is not blind chance. It's only blind chance if you start with the materialist dogma that it *has* to be blind chance. Alfred Russel Wallace, who, together with Charles Darwin, discovered the principle of natural selection and founded evolutionary theory, ended up with the idea that evolution is guided by intelligent spirits, that the creative side of evolution is guided by an immanent creative intelligence, or many kinds of intelligences, within the natural world. And that's just as compatible with the evolutionary facts as the neo-Darwinian dogmas. However, even if evolution is guided by intelligent spirits or by intelligence immanent in nature, that doesn't necessarily mean that this intelligence is working in accordance with an overall master plan—or that human cultural evolution is guided by an intelligence immanent in human beings. Every innovation, every gadget that's invented, every new advertising slogan, every new piece of music or work of art that's made is guided by a creative intelligence. But that doesn't mean that we know where we are going. It doesn't mean that these creative intelligences are working in accordance with some master plan for the

> destiny of humanity. Mostly, they are working in accordance with much more short-term goals.
>
> For me, it's an open question as to whether the intelligence that underlies the creativity in life is working in accordance with some fixed goal for the end of evolution. I don't get that impression. If you look at the diversity of life—several million species of beetles, for example, on this planet—you get the impression that there's a kind of creativity for its own sake, a proliferation of form and variety. It's not at all clear why there should be so many millions of species of beetles.
>
> Only in the Jewish religion, and in derivative religions like Islam and Christianity, do you have this very strong emphasis on process and time. And now the West comes up with evolutionary theory, and suddenly it turns out that this is the process not just of biological life on earth but of the entire universe. Is this a vast cultural projection and justification of our religious assumptions? Or is it a fantastic confirmation of them from science. It's hard to know which.

We started by wondering if speaking of Sufism and modern science made sense, and if so what the relation could be. I have tried to point out affinities in thoughts and intuitions, "correspondences" that seem to create bridges across the centuries, connecting cultures and worldviews at first sight utterly alien to each other.

Here is one more correspondence: perhaps Teilhard de Chardin may provide an answer:

> By means of all created things, without exception, the divine assails us, penetrates us and moulds us. We imagined *it* as distant and inaccessible, whereas we live steeped in its burning layers . . .
>
> God reveals Himself everywhere, beneath our groping efforts, as a *universal milieu*, only because He is the *ultimate point* on which all realities converge. Each element of the world, whatever it may be, only subsists, *hic et nunc*, in the manner of a cone whose generatrices meet in God who draws them together (meeting at the term of their individual perfection and at the term of the general perfection of the world which contains them). It follows that all created things, every one of them, cannot be looked at, in their nature and action, without the same reality being found in their innermost being—like sunlight in the fragments of a broken mirror—one beneath its multiplicity, unattainable beneath *its* proximity, and spiritual beneath its materiality. No object can influence

us by its essence without being touched by the radiance of the focus of the universe.[24]

Rumi or Jami couldn't have expressed it better in their verse, I think.

Notes

1. Pierre Teilhard de Chardin, "The Evolution of Chastity" and "The Sense of Man" in *Toward the Future* (London: Collins, 1975).
2. Virginia Del Re McWeeny, *Persia Mystica, Poeti sufi dell'età classica* (Pisa: Edizioni ETS, serie Philosophica, 2004).
3. To mention only a few: Qâdiriyya, active mainly in the Middle East; Naqshbandiyya in the Middle and Far East; Chistiyya, in the Indian subcontinent; Shâdhiliyya, in North Africa and the Middle East. In Italy, the Jerrahi al Halveti of Genoa is a well established Sufi tekke (Order): one of its objectives is the Christian-Islamic interreligious dialogue.
4. Musician, initiated at the school of Nizani—a branch of the sufi order Chistiyya, founded by Nizam-ud-Din Aulia (1238-1325).
5. *The Mathnawi of Jalalu'ddin Rumi*, 8 vols, edited with critical notes, translation and commentary by Nicholson, R.A., Brill, E.J., Leiden-London 1925-40.
6. Translated by N. Pourjavady as *Sawanih: Inspirations from the World of Pure Spirits, the Oldest Persian Sufi Treatise on Love* (London and New York: KPI in association with Iran University Press, 1986).
7. Jami, Mulla 'Abd ar Rahman, *Masnavi-yi-Haft Aurang*, ed. by Mudarris Gilani, Tehran, 1958, transl. J.T.P. De Bruijn, in *Persian Sufi Poetry* (Richmond, Surrey: Curzon Press, 1997).
8. In: Andrew Harvey and Eryk Hanut, *The Perfume of the Desert* (Adyar, Madras, India, Quest Books, Imprint of Theosophical Publishing House, 1999).
9. In: Nicholson, R.A., *Translations of Eastern Poetry and Prose* (Cambridge: Cambridge University Press, 1922).
10. Physicist and philosopher, a disciple and friend of David Bohm, is the author of many books on science as well as creativity, synchronicity, Jungian theory, science and art, and non European cultures. He is founder and director of the "Pari Center for New Learning ", www. paricenter.com in the village of Pari, between Siena and Grosseto, Tuscany, where he lives with his family.

11. Henry Corbin, *The Man of Light in Iranian Sufism*, transl. from the French by Nancy Pearson (New York: Omega Publications, 1994). Ch. V. "The Black Light"; § 1. "Light Without Matter", p. 100.
12. "Science and Sufism", by William C. Gough, President: Foundation for Mind-Being Research. Published in the Proceedings of the Fourteenth International Conference on the Study of Shamanism and Alternate Modes of Healing, held at the Santa Sabina Center, San Rafael, CA, Aug 30-Sept. 2, 1997.
13. Annemarie Schimmel, *Sufismus, Eine Einführung intdie islamiche Mystik* (München: C.H. Beck Verlag, 2000). A. Schimmel is the author of many famous studies on Rumi's poetry and imagery, as well as on aspects of Iranian Sufism.
14. From Hal Blacker's interview with Rupert Sheldrake, on the occasion of the publication of *The Physics of Angels*, by Matthew Fox and Rupert Sheldrake (San Francisco: Harper, 1996, online on http://www.wie.org/j11/sheldrake.asp). See also: R. Sheldrake: *New Science of Life* (Rochester, NY: Inner Traditions, Bear & Company, 1981); *Seven Experiments that Could Change the World* (Rochester, NY: Inner Traditions, Bear & Company, 2002); also Rupert Sheldrake, Terence McKenna and Ralph Abraham, *The Evolutionary Mind*, (Rhineback, NY: Monkfish Publishers, 2001). David Peat mentions Sheldrake in his autobiography, *Pathways of Chance*, Pari Publishing; Italian tr., *I sentieri del caso* (Roma: Di Renzo Editore, 2004, p. 96) and in *The Blackwinged Night* (Cambridge, MA: Perseus Publishing, 2000), p. 163.
15. Hal Blacker, ibid.
16. A. Zee, *Fearful Symmetry: The Search for Beauty in Modern Physics* (Princeton, NJ: Princeton University Press, 1999).
17. Mathnavi, vv. 1960-1968 (Nicholson ed.). See ref. 5.
18. F.C. Happold, *Mysticism, A Study and an Anthology*, Pelican Books, 1970; reprinted Penguin Books, Middlesex, England, 1990.
19. op.cit, p. 31
20. In: E.G. Browne, *A Persian Anthology*, 1927 (quoted from Mathnavi, IV lines 2881, Nicholson, ed. op.cit.). The final line refers to the motif of the return of all human being to Adam, to the positive non-being.
21. Happold, op. cit.
22. Happold, op.cit.
23. Happold, op.cit.
24. Teilhard de Chardin, *The Divine Milieu* (London: Collins, 1960).

Colin Tudge

Colin Tudge *was born in London in 1943, educated at Peterhouse, Cambridge and was features editor for* New Scientist *magazine, before joining the BBC where he presented* Spectrum. *He is a regular contributor to newspapers and magazines including* The Independent, The Times, Natural History *and the* New Statesman. *He is a former member of the Council for the Zoological Society of London and is visiting Research Fellow of the Centre for Philosophy at the London School of Economics. He has a lifelong interest in biology and a long-standing interest in farming, food politics, "various bits of philosophy", and is especially interested these days in the relationship between science and religion—"both are necessary". His most recent book,* Feeding People is Easy, *is published by Pari Publishing.*

SCIENCE, RELIGION, AND ME PERSONALLY

COLIN TUDGE

My involvement with religion begins with the simple acknowledgement that the world is not what it seems: that, and a feeling that it is highly desirable and might be possible during our brief stay on this Earth to get beyond what seems, and to acquire a greater sense of what really is, and of how and why things are as they are.

Many of course would claim that this is precisely what science does—contrives to improve on common observation and on common sense. Indeed it does and to an impressive extent it succeeds. $E=mc^2$ is surely a wondrous insight and there is nothing obvious about it—it certainly gets beyond what "seems". But although science is obviously wonderful it is equally obviously limited: indeed it succeeds as well as it does largely (or indeed entirely) because it limits its own ambition. As the great 20th century zoologist Sir Peter Medawar put the matter, "Science is the art of the soluble". Scientists tackle only those things that they can observe, and measure, and test. That leaves a great deal out. Indeed, it even leaves out a great many issues that seem in principle to be within its bailiwick—such as whether there is only one universe, or a great many, or indeed a potential infinity of universes: the "multiverse" idea. This matter is intriguing, and important, and has to do with real physical dilemmas—there is none more real—and yet it is beyond experimentation. Scientists cannot explore what lies beyond our own universe. They can only speculate that there might be more than can be directly experienced. Matters of a non-material nature, that have to do with love, or beauty, or right and wrong, or indeed with the meaning of "meaning", are outside the scope of science altogether. But they seem important, too.

Many scientists know full well that science is limited but some nonetheless insist, whether they believe it or not, that science and logic are all that is worth taking seriously. The most rigorous of these hard liners argue as the logical positivists did in early 20th-century Vienna, that all questions that *cannot* be approached simply through observation, measurement, and logic—the three main struts of "reason"—were and are, literally, meaningless. But although this positivist rigour may in some sense be admirable (rigour is always admirable up to a point) it also seems defeatist. It implies that whatever we cannot apply our slide-rules to must forever be beyond out ken, or indeed beyond worthwhile contemplation. Such rigour is also dehumanising. The logical positivists seem to want us to think in the way that modern computers do. Yet one of the greatest insights of AI (artificial intelligence) in the first 30 years or so of its formal existence is that human beings are *not* computers: or at least, we do not belong to any generation of computers that is yet conceivable. We think more vaguely than computers do, we make more mistakes, but we also perceive and grasp—"prehend", was A.N. Whitehead's much-neglected term—a great deal that computers do not, and perhaps never can. Computers are wonderful as science as a whole is wonderful but they are also limited as science as a whole is limited. Both deal in certainties only because they are expressly designed to explore those issues that *can* be resolved with certainty. It is a privilege to be born a human being, with three and a half billion years of evolution behind us, the direct progeny of the universe itself. With such advantages it seems perverse to contrive, deliberately and as a matter of policy, to think like a mere machine, albeit a very clever machine. I am sure we can do better than that—or at least it's worth making the effort. The noun that derives from "prehend" is prehension. Religion, as I see it, is an exercise in prehension: grasping all that it is within our power to grasp, with no a priori restrictions on what might be achieved.

But although we may not be content simply to think like computers, or to suggest that science offers the only insights that are worthwhile, this doesn't mean—does it?—that we need to be religious. Artists operate in spheres that lie outside science. In general they contrive to express their emotional responses to the universe around them. This is an important thing to do—we are right to take art seriously—and, clearly, it is not science. Yet art *per se* is not religion *per se*. Many artists are vehement atheists—or at least they say they are, and it seems presumptuous to doubt their word. Humanists deal with all aspects of human affairs, including those of morality, but many of them make a virtue of their atheism.

But I'm afraid that at this point I tend to shift the goalposts. I do not want to define religion in the usual ways. I don't want to say as many do that religion is inveterately concerned with the "supernatural". I certainly don't want to say that it necessarily involves any particular God who can be named, and who must be worshipped via particular rituals and ceremonies. I am all for rituals and ceremonies—but I see them as psychological and social devices, ways of focusing the mind; not as metalled roads into the presence of God; nor, indeed, as magic.

In short, I do not suppose that you have to be a paid-up Christian or a Muslim or a Jew, or a Hindu or a Buddhist or a Sikh or a Jain, to be religious. Each and all of these disciplines may help the process of prehension—the formalism can be helpful, and the gathering of like-thinking people on a difficult and common quest certainly is—but no single, discrete "faith" encompasses all that is meant by religion. Each is merely an expression of the religious quest, the desire to prehend, to grasp all that can be grasped (and then some more), and each offers a way of marshalling thoughts and feelings.

Some traditions in some religions seem explicitly to endorse this way of looking at things, at least to an extent. The prophet Mohammed warned that it is blasphemous to try too hard to define God. God is always well beyond our understanding. Moses is at the root of the Judaeo-Christian tradition: and as he approached God on the mountain (Exodus 3) he was greeted in the end, only by light. He could not see beyond the light to the reality behind. The idea of God, in short, should in some sense be left vague. Vague does not as some atheists might triumphantly insist, mean woolly. It does mean a sense, a recognition, that at the heart and the root of everything, universe or multiverse, is mystery. The word "ineffable" applies: that which cannot be expressed. Hindus insist that all religions, sincerely undertaken, are paths to God. Buddhists have rarefied the idea of God to the point where in effect there isn't one—but they still talk about the universe as if there was. Buddhists have a very strong sense of God although, in sharp contrast to Jews and Christians in particular, they have no inclination at all to delineate God, or to attempt a definition.

I suggest as a matter of simple empirical observation that what all the formal religions aspire to do—not only the ones already alluded to but also those of the Maoris and the Hopi and a hundred thousand other peoples from the past and present—is to provide a complete account, an all-embracing narrative, of all that is and all that concerns us. What the universe is like; how it came into being; what human beings are; how we should behave—these emerge as the outstanding questions within those narratives. All great religious texts, including the Bible and the Qur'an, deal with all of these issues. I stress the word "narrative". For the point is not merely to list all possible issues in a series of essays, as in an encyclopaedia. All the strands must cohere in one grand story. I am not at this point saying that this is what religion *should* do, or that this is how religion ought to be defined. I am merely suggesting, as a testable observation, that this is what religions in practice do. All the religious texts that are seen truly to be seminal are all-embracing narratives, into which everything fits in one grand sweep. If "religion" is to be defined at all, it seems to me at least reasonable to begin with the fact of what they actually do.

The grand narrative is invariably generated by a collective. This is obviously true of the Bible, the work of many scholars and prophets over many centuries. Islam and Buddhism are each perceived to be the creations of single individuals—but Mohammed acknowledges his debt to the great Judaeo-Christian prophets and Buddha inherited the broad tradition generally known as Hindu. Both those great founders might have acknowledged, as Isaac Newton did in a different context and in a different age, that they had stood on the shoulders of giants. Both Islam and Buddhism have been developed in various ways since their founders' time, by generations of scholars and devotees. In short: the narrative that each religion provides is always the work of entire societies and traditions, even when it seems to be the work of one prophet. The author of religion as a whole, is all humankind. The all-embracing narrative is the collected experience and perception of humanity.

If you think of religion as the all-encompassing story then many of the traditional arguments that surround it disappear. In particular—the main theme of this essay—what is the relationship between religion and science? On the spectrum of current views on this three particular positions stand out.

The first, embraced by out-and-out atheists such as Richard Dawkins and Jonathan Miller, sees religion as obsolete. Religion was and remains a primitive way of looking at the world—one that sought to explain the way things are simply by suggesting that some outside agency, some genderless super-being known as "God", made them that way. This crude attempt at explanation has been replaced, say Dawkins and Miller et al., by science. Charles Darwin's theory of evolution by means of natural selection serves these days as the test case. Genesis, the first book of the Bible, and so the first text of Judaism, Christianity, and Islam, says that God made the world and all the creatures within it in six days (and rested on the seventh). God made them, it implies, as they are now. Evolutionary theory says that in reality living creatures took billions of years to emerge; that they changed over time ("descent with modification" was the expression that Darwin preferred to "evolution"); and that they have been shaped since their beginning not by the guiding, benign or not-so-benign hand of God, but by "natural forces": that is, by natural selection. According to Dawkins (and many other biologists) the Genesis account is simply wrong—factually. Dawkins' own account of how living creatures came to be as they are, including humans, is entirely mechanistic. Once you understand the mechanism—Darwinian natural selection—then, he says, you see that nothing else is necessary. All the rest is just fairy-tale and wishful thinking (or plain wrong).

The second common view of science and religion, embraced by people known as "fundamentalists" (who are more likely to be Jewish, Christian, or Islamic than Buddhist or Hindu), takes a particular religious text as the literal word of the one true God and sees any deviation from it as blasphemy. Blasphemy in turn is commonly perceived as a threat to us all and so it can and sometimes should be punished by the most extreme measures. Darwin's *On the origin of species by means of natural selection* has often been banned and sometimes burned and those who espouse it have often been deprived of their livelihood and ostracised, and sometimes beaten up even if never, to my knowledge, publicly executed. Although many fundamentalists make free and even lavish use of the technologies that emerge from science—the so-called "high" technologies, such as 4x4s and mobile phones, not to mention plastic surgery and the other fruits of modern medicine—many tend to be suspicious of science as a whole. They are not comfortable with its overall agenda. They are inclined to feel that any attempt to probe the mysteries of the universe is somehow blasphemous.

The third common view of the relationship between science and religion is that they simply have different agendas. Stephen Jay Gould expressed this idea in a recent book (*The Hedgehog, the Fox, and the Magister's Pox*, Vintage, 2004). Science, he said in effect, answers "how" questions; and religion addresses "why" questions. The two are not in conflict. Indeed they do not even overlap. They complement each other.

But if you take the view, as I do, that religion is and aspires to be the all-embracing narrative, the complete account of all that is and could and should be, then none of these common positions will do. Religion, the way I see it, embraces all formal disciplines, including science. In this view, followers of the three "Abrahamic" religions that have taken their lead from the Bible should delight in good science, including the evolutionary ideas of Darwin. Some suggest that the authors of Genesis were attempting to describe how the world literally came into being—and given their almost total absence of formal data they didn't make a bad fist of it. The emergence of the Earth from the void, the separation of the waters, and the gradual step-by-step accumulation of living creatures matches what are now construed to be the facts commendably well. Others suggest that Genesis should not be read as a literal account, but as a metaphor. Some, like the popular theologian Karen Armstrong, suggest that *all* religion is metaphor—while emphasising the supreme importance of metaphor in all human understanding. I like both these points of view. In either case, I see no reason why someone who likes the literal idea of the Creation should not be grateful to Darwin for explaining so cogently how God might have performed his wonders. Natural selection is really rather clever, after all, and tremendously effective. Many a Christian priest and theologian has taken this line.

The same principle can be applied to science as a whole. Religion is concerned with the facts of the case, as well as with everything else. Science, beyond doubt, extends our insight into the facts. If religion is perceived as the complete narrative, with all humanity as its author, then the insights and precision of science improve it: make it stronger and deeper. This way of thinking is in line with that of the great 17th-century founders of modern science: Galileo, Descartes, Leibniz, Newton, Boyle, John Ray. All were devout, and all saw their science as an aspect of their religion and as a way of reinforcing it—the proper use of the God-given intellect, the better to understand God's works. Research, for them, was an act of devotion. I don't see what's wrong with this. Indeed I like in this context to misquote Marx. He suggested that the proper role of philosophy is not to understand the world, but to change it. I suggest, as John Ray might have done, that the proper role of science is not to change the world, to provide technologies to make us more comfortable, but to deepen our appreciation of it. The idea that it is blasphemous to seek to explain the wonders of the universe seems itself to be blasphemous. God is not a conjuror, whose tricks seem tawdry once exposed. It is as Socrates (and many another thinker) has discovered: the more that's known, the more wondrous it all seems.

Some may have baulked at the idea that religion is "narrative". On the one hand the great religious texts (especially the Bible) are compounded in large part from stories of particular people and events, so the idea that they collectively form a narrative might seem perfectly reasonable. On the other hand, "narrative" implies "story", and many religious people might be offended by the idea that their faith is a "mere" story. This needs teasing out.

Religion and science as narrative

The connotations of the term "narrative" run far deeper than is obvious. Narrative does indeed imply "story"—but that is no kind of insult. Whether we prefer to say that human minds are evolved, or that they are designed, the fact seems to be that they are very good indeed at remembering and spinning stories, especially stories with people in them, and deal much less convincingly with lists or abstractions. Aristotle saw tragedy as the highest form of literature and declared that above all it must have a beginning, a middle, and an end. It must be coherent and have a denouement. Our way of thinking contrasts sharply with that of computers, or at least with computers so far. Computers are very good at lists—no problem at all—and they seem to have very little use for the niceties of literature. Sub-plots for computers are a distraction. For us they are a useful *aide-memoire*: "secondary cues", the psychologists say. Mnemonics are a simple example of the principle.

The sum total of all that we have in our heads, our knowledge and our attitudes, may be called our "worldview". And I suggest that our worldview takes the form of a narrative. It is a commentary on the world around us. It is a story that we tell ourselves. Some parts will surely be true—in the sense that they do indeed describe, reasonably well, some aspect of the universe. Some will be padding—forms of words and ideas that do not necessarily correspond directly to anything that's out there, but which serve to hold all our information and emotional responses together, in one coherent whole. When we say that we "understand" something what we really mean is that we can comfortably fit it in to our own personal commentary on the world; that it can take its place in our own private story. The things that we hold to be "true" may or may not be literally true. But we take them to be true because the particular story they tell us seems to fit in neatly with what we already take to be true, and perhaps plugs a conceptual gap. But in the end, for each of us, truth is a story that we tell ourselves that for the time being happens to seem convincing.

It is in no way insulting to suggest that religion is narrative. *All* our understanding of everything is narrative. Everything we think we know and understand is our own private story, unfolding in our heads. Some of it—much of it—is surely true. Some of our personal story may be inherited. Some of it may be learned. Some of it may, as prophets claim, be revealed to us by God. Whatever the source, all in the end is knitted neatly into the grand fabric of our personal account of how things are. That's the way our brains work. It is wonderfully economical.

Science is narrative, too. It is not typically seen or presented as such. Its advocates tend to present it as an edifice, a great monument of truth, made up of facts that are unequivocal and theories that have been tested and proven beyond question; a citadel that builds inexorably stone by stone and one day, perhaps soon, will contain all that can be known or is worth knowing.

But science is not like that at all. No facts beyond the most banal—facts that are merely matters of definition, as in "grass is green"—can ever be unequivocal. Theories cannot be "proven" in the sense that we can ever be sure that they fulfil the lawyers' criterion: the truth, the whole truth, and nothing but the truth. Certainly not the whole truth. Medawar's rule applies: scientists explore only what it is in their power to explore, and what it occurs to them to explore. Their most sweeping and tested theories deal only with aspects of the whole. The inverse rule applies: the broader the sweep of the theory, the less certain it becomes. The further we wander from facts that are banal and are simply matters of definition, the more the uncertainties creep in. More certain must mean more trivial. The logical positivists and other philosophers of the early 20th century conceived that everything that there is could be explained and expressed, in the end, in mathematical terms. Then along came Kurt Gödel who showed that maths itself is uncertain. All but the very simplest mathematical expressions are bound to contain terms or reasoning that are arbitrary, or else the theorem breaks down. Time after time in science we have seen theories and methods arising that could be applied to ancient dilemmas like algorithms: let the method loose and it will tell us the answer once and for all. Well, some problems do yield to such crunching but the bigger problems decidedly do not. I am interested in taxonomy, the classification of living things. In the 1980s, the new discipline of cladistics combined with detailed studies of DNA was supposed to show once and for all which creature was related to which and who was whose ancestor. But in reality the data are full of ambiguities, and the conclusions drawn from them depend on which statistics are employed for their analysis, and these matters must be decided by human judgement, just as was the case in the days of Richard Owen in the early 19th century when all there was to go on was bones. There are no royal roads to truth.

In short, the corpus of science is not an edifice. It is more like a landscape painting rendered in oils. It is built up over centuries—science takes an awfully long time—brushstroke by brushstroke, and any one stroke is always likely to wreck the balance of the whole and require the artist to start again—the "artist" in this case being the scientists of any one particular discipline. Landscape paintings are satisfying only if they fulfil the criteria of aesthetics. They must be nicely balanced and also, that word again, "coherent". But despite John Keats' assurance that "Beauty is truth, truth beauty", aesthetics can be highly deceptive. "Beauty lulls", as C.S. Lewis said. In any case, the landscape painting of the universe as a whole that physicists now present us with is far from coherent. The "classical" theories that run from Newton's mechanics through to Einstein's relativity simply do not square up with those of quantum mechanics. Physics currently presents us with two incompatible worldviews. The picture it provides of the universe is not a classical landscape, in which we can all wallow. It is like the cubist vision of Braque and Picasso: a pot-pourri of fragments;

snapshots. So, in reality, is all science. It's just that we usually don't notice. But the fragmentation of physics, perceived to be the most fundamental of all sciences, is all too conspicuous.

I love science. For me its uncertainty is part of its charm. I would hate to think that it could ever get to the end of whatever roads it happens to be on. But we should recognise its shortcomings. It is wonderful, and does wonderful things. But we shouldn't get carried away by it, or put too much store by it. Its view of the universe will always be uncertain, and always fragmented. Only in very special contexts, *faute de mieux*, should it be perceived as the final arbiter.

But let's look in more detail at the narrative of religion—the broadest of all narratives, in which that of science is embedded. What's in it?

The narrative of religion

If, as I claim, religion is humanity's grand narrative then it must contain a virtual infinity of sub-plots. Nothing escapes it. But in practice those plots fall under three headings. Some deal with what the universe is really like and how it came into being: this is Cosmology. Some deal with matters of Morality: right and wrong; how we should behave and with what motives. The third might loosely be termed Epistemology: how we can find out what is true; how we know truth when we see it.

The cosmology of religion should embrace science. "Render unto Caesar that which is Caesar's": it is simply foolish to insist that the facts as related 3000 years ago by scholars who had no telescopes or microscopes and had done very little travelling should be matched against modern findings. Anything more than a superficial reading of the Catholic authorities' dispute with Galileo shows that in the beginning at least, before the arguments became political and acrimonious, the Catholic theologians acknowledged full well that this was the case (and Galileo was a good Catholic too).

But the agenda of religious cosmology goes beyond science, and so it should. Science can do only what science can do; and if we don't go beyond it, even within what it claims as its own agenda, then our insight must be limited. It is a shame, an exercise in self-immolation, not to pursue ideas where they may lead, even if they do lead us beyond the bounds of experimentation.

Two modern examples are particularly striking. The first is already alluded to: the idea that the Universe as now observed may not be all that there is. The problem is that the Universe would not hold together—there would be no atoms, and so there would be nothing at all that we could recognise as "substance", and of course there would be no "we" to do the recognising—if the physical laws and constants that contain the universe were not so finely tuned. If the proton was ever so slightly bigger and heavier, the entire universe would collapse on itself. So it would too if gravity was slightly more forceful. And so on. Physicists calculate that the chances that all the constants of the universe should be as precise as they are—and as they need to be if the universe is to function—are about 1^{-120}: one in trillions of trillions of trillions. There are only two logical ways out of this dilemma, they argue. Either the universe is intelligently designed—set up by some outside agency and fine-tuned all its physical details: an outside agency who by definition is God. Or there is, in reality, a potential infinity of universes—a "Multiverse"—that are then subject to the inexorable "law" of natural selection. Only those universes that do have nicely balanced constants survive; and our own particular Universe happens to be one of them.

There is a neatness about the Multiverse argument that appeals to many scientists and mathematicians. It might have particular appeal for those who are atheist. After all, it seems to explain why the present Universe is as finely-tuned as it is even without recourse to an intelligent designer, the idea of which they find abhorrent.

But there are two great snags. The first is that science qua science can and should deal only with what is observable and measurable, and can be tested. But, by definition, the Universe is the sum total of all that can be observed. So in positing other universes, scientists are going beyond the realms of science. What can they do with such a problem? Just stop thinking about it? Or acknowledge that it ought to be thought about—and yet is beyond direct investigation of the kind that is the *métier* and the *raison d'être* of science?

The other snag was discussed very thoroughly by Thomas Aquinas in the 13th century. The wondrous thing, he said, the thing that really needs to be explained, is not how things happened, but *how come* they happened. Genesis tells us that the world arose out of a void: that something came of nothing. How is that possible? Why should such a thing have happened? What is or was the nothingness out of which the universe emerged? How did it get there? Ludwig Wittgenstein made the same point 700 years later in his *Tractatus Logico-Philosophicus*: "Not how the world is, is the mystical, but that it is". The word "mystical" is strange in this context but the meaning seems clear. It isn't enough simply to pronounce that such and such-a-thing happened in such a way. Truly to understand, we need to know what was going on before the thing happened that made it possible, and how it could have come about. Scientists might choose to say, in the manner of logical positivists, "The question cannot be answered and so is meaningless". But big questions don't go away just because some people decide not to take them seriously.

My second example of a seriously big idea that profoundly affects the physical world and yet is beyond conventional science is that of mind. Mind, in the end, is all we have. Mind contains our personal commentary on the universe. In effect, for each of us, it is the universe. For each of us, and for humanity and all living things as a whole, it is the most important thing there is. But what is it exactly?

René Descartes, in the 17th century, was clear in this as in all things. The Universe has two components: mind, and matter. They are separate, although each influences the other. This is "dualism". A century or so after Descartes, the Irish cleric Bishop Berkeley declared that mind is all there really is. Matter is an illusion. Some modern philosophers, including the American Daniel Dennett, have taken the completely opposite view: that matter is all there is. Mind is the illusion. At least: we come aware of "mind" through our consciousness, and consciousness is merely an "emergent property": just the feeling we get when we think. It is not quite clear to me why this kind of argument is supposed to be an explanation. But some people are clearly convinced by it, and what people are convinced by they construe to be the truth.

But there is a fourth kind of argument: one that arose out of science in the early 20th century and has been powerfully reinforced ever since. It properly began with the proposal of one of the earliest modern quantum physicists, Eugene Wigner, that the outcome of experiments carried out on fundamental particles was profoundly influenced by the thoughts of the experimenters themselves (the Schrödinger Cat paradox). This idea has never gone away. In its strongest form it emerges as the "anthropic principle": the idea that the Universe as a whole could not exist unless conscious minds are thinking about it (and it becomes "anthropic" because we, anthropoids, are the only conscious creatures of whom we are aware).

I have no great love for the anthropic principle but I very much like a related idea: that the Universe is not made of matter plus mind, as Descartes suggested; or of mind alone, as Berkeley posited: or of matter alone, as Dennett and his fellow hard-liners insist. The basic stuff, I suggest, might be called "mind-matter". The idea that the most basic stuff seems to be a compound of two different things rather than a single thing should not alarm us. After all, common sense tells us that there is something called space, and something called time, and that each are fundamental. Newton, common sensically, treated them as separate entities, and the physics he built on the back of that common sense was very successful. But Einstein showed, now beyond reasonable doubt, that time and space are interdependent. The most fundamental quality is neither space nor time, but "space-time". At least by analogy, it seems to me entirely unexceptionable to suggest that the basic stuff of the Universe is neither mind nor matter, but mind-matter. Human beings and other mindful creatures are conscious insofar as they partake of, and consolidate and magnify, the mind that is part of the fabric of the universe, and so is all around us.

I derived this notion from ideas of the British particle physicist Chris Isham. He may prefer to disown my version of it. But he did make a general comment that seems to me absolutely to the point: that whatever the nature of mind, and its relationship to matter, we are, at present, "A million miles from understanding it". The attempts of some scientists simply to banish the issue of mind from their discussions seems to me to be merely uninteresting, except in what it tells us about their psychology. The suggestion of philosophers like Dennett that everything can be explained already is extraordinarily pre-emptive and premature; and it is bad philosophy, it seems to me, to pretend that a subject is sewn up even before it has been seriously engaged.

Again, though, we find that the mind-matter idea, and all serious thought on mind, stretches science to breaking point, and probably beyond. It's not simply that present-day science is not up to the task. It may be rather that the questions posed are simply not within the brief of science. To explore them, we have to indulge in cogitations of a kind that are not scientific. If you decree that the cogitations of science are the only ones worth indulging, then you drop out of the debate. That is what some scientists and philosophers have chosen to do and apparently would like us all to do. But it is not obvious to me why we should.

If the mind-matter idea does stand up, if indeed it is true, then it changes everything. Its implications for some of the core ideas of religion need thinking through but they seem intuitively obvious. It re-admits the idea of the intelligent Universe: the Universe that "knows" what it wants to be. It throws new light on the ideas of thinkers such as A.N. Whitehead and Teilhard de Chardin. More broadly, it leads again into the idea of the immanent God (the God who is ever-present in the Universe), as opposed to the transcendant god (its detached Creator and arbiter). It suggests that prayer, which to some extent may be seen as an attempt to influence events by thinking, might have a physical basis. More broadly, it gives new substance to the idea of mysticism. Mysticism becomes the endeavour to engage the mind that is within our own heads directly with the mind that is in the universe as a whole.

The second thread in the grand religious narrative is moral: how we can identify what is good. Moral philosophers distinguish three main approaches: utilitarian; "absolutist"; and "virtue ethics". Utilitarianism was summarised and largely formulated by the late 18th-century British philosopher Jeremy Bentham as "the greatest happiness of the greatest number". Goodness and badness are judged purely by outcome, and outcome is measured in human happiness. But this is clearly inadequate. If the greatest happiness of the greatest number is all that matters, why isn't gang-rape justified? A lot of men are made happy, and only one person misses out. No sane person would defend such an argument, but to argue against it we have to invoke principles beyond mere hedonism and the counting of heads. What are those principles? Where do they come from? Immediately we need more criteria by which to judge.

The ethics that emerges from religion tends to be absolutist: what is good is what God says is good. God is perceived as the source of absolute moral Law. But this raises two main kinds of problems. The first is to find out what God is actually saying, and whether the God who seems to be issuing orders is or is not the true God. Perhaps he is some false god, or demon, trying to fool us. In practice we rely on prophets to tell us what the true God actually wants—and we pick our prophets. The ones who have survived to tell the tale seem on the whole to be saying good things. But still they might be fooling us.

The other theoretical problem is the one raised both by Immanuel Kant and then in a somewhat different form by Friedrich Nietzsche. There may be a good that is greater even than God. God may be capable of behaving badly. Certainly, some of His behaviour as described in the Old Testament in particular can seem equivocal. His insistence that Abraham should sacrifice his own son, his torment of Job, and his constant exhortations to smite and destroy the various enemies of Israel, seem gratuitously cruel, even if (sometimes) they did turn out for the best in the end.

But in practice religions do not deal exclusively or even primarily in law. To a striking extent both testaments of the Bible deal in virtue ethics. The key to virtue ethics seems to me to lie with David Hume who, rationalist that he was, pointed out that reason alone cannot lead us to moral certainty (we cannot derive "ought" from "is"). The root of all morality lies in the "passions", he said: emotional response. This, I suggest, is the prime focus of religion. Religion cultivates and refines appropriate emotional response. Prophets most obviously differ from secular moral philosophers in their emphasis, not on particular lines of action—not on particular "laws"—but on attitude. The Ten Commandments contain some specific prescriptions—no killing, no adultery—but mostly they focus on attitude: reverence for God, respect for parents. Mohammed emphasised humility in the face of Allah, and generosity to the poor. Christ emphasised universal love. Buddha spoke of compassion. The Hindu mystic Ramakrishna boiled all the moral imperatives of all the great prophets down to three: personal humility; respect for other sentient beings; and reverence for God (although pantheists might equate "God" with the universe as a whole). This to me seems beyond improvement—and cuts a remarkably wide swathe through modern debates on, for example, human cloning or Third World development or what you will.

The religious prophets do not have a monopoly on virtue ethics: *vide* Aristotle, Lao Tsu, and many Romans. But we could argue this point the other way around. Aristotle was the great rationalist. But insofar as he argued virtue ethics, he was arguing religiously.

This brings us neatly to the third prime leitmotiv of religion: epistemology; how we find out what is true, and know truth when we see it. Those who are both scientists and atheists are clear. All ideas that are worth admitting must be rooted in what can be directly and repeatedly observed and preferably measured and subject to thought of the strictly circumscribed kind known as logical. Preferably—even essentially—the ideas should be formulated and manipulated in the symbols of mathematics, in which the logic is clear to see and is inexorable. Scientific papers reflect this rigour—or at least, they are supposed to. Every statement is backed up by transparent data, rigorous reasoning, and preferably by a stack of references.

Ordinary people don't think this way. What most of us believe is a mish-mash of unexplored impressions, vague memories, and hearsay—as well as the ideas that we have learned formally, and passed exams in, and which we have put into practice in our own lives and found for our own purposes to be sound.

At first sight, the method of science seems far more reliable. It is as near to being chanceless as can be devised. Yet as emphasised throughout this essay, the insights of science are inevitably limited and to a large extent are self-deceiving. Its truths cannot be certain, and such certainty as may exist is attained by limiting the terms of reference—cutting out everything that is too hard to measure. That doesn't seem very satisfactory. It certainly doesn't look like a complete formula for living.

Religion doesn't simply do what most of us do through most of our lives. It doesn't simply accept whatever ideas happen to have floated its way, or seize upon whatever is convenient. Neither, as the common cliché has it, is religion based purely on faith, and neither, as atheists would have us believe, is faith nothing more than a perverse but self-serving desire to believe the implausible.

Emphatically nor does religion properly defined eschew reason, the commodity that many scientists and philosophers would like to claim as their own. The best theologians are unremittingly rigorous. No-one has ever thought more deeply and precisely than the great scholars of the Middle Ages. Not for nothing was Duns Scotus known as "the subtle doctor". Aquinas spent much of his long life seeking to reconcile the rationality and science of Aristotle with the teachings of Christianity. It was not possible, he concluded, to "prove" the existence of God by reason alone but it was possible, and necessary, to move by a series of eminently rational steps along the path that led to faith. In the following (14th) century the English Franciscan William of Ockham took an almost opposite view. Reason is reason, he said, and faith is faith, and the gap between them could not be bridged. His own power of reasoning was wonderful. He founded some of the key principles of modern science. But he stressed, nonetheless, the supremacy of faith.

Faith remains an elusive concept. In the end, Ockham is surely right. There is a conceptually unbridgeable gap between reason and faith. You cannot argue your way step by logical step from atheist materialism to a belief in God. Yet it is not necessary, as is so often suggested both by religious people and by atheists, simply to make a blind "leap" into faith. In any case, faith is a long way down the line. Roman Catholics speak of the "gift" of faith. Some people—St Paul is the outstanding example—have experienced instant conversion apparently from one extreme to the other. But others have entered religion by slow immersion: C.S. Lewis processed from atheism to Christianity reluctantly but absolutely, largely through long conversations with his friend J.R.R. Tolkien. Buddha achieved enlightenment in one moment of mysticism under a bo tree—but only after years of Hindu practice. The first step is not to "leap" but simply to take religion seriously. This, at the very least, it deserves. My main beef with the atheists who currently occupy British TV, including Dawkins and Miller, is that they do not engage with the subject. Neither do they seem to have tried to do so. I suppose I could appear on television and rail against ice hockey, which from my position as an extreme outsider seems most unpleasing; but if I was invited to do so, I think I would at least take the precaution of learning the rules first. Anyone who seriously tried to understand what religion's many prophets and theologians are actually saying is likely to find at least some of it intriguing, and after that who knows? To this extent Aquinas is surely right. Reason alone cannot take you to faith, but it can certainly help to prepare the ground—and it is necessary. But my point here is not to promote or to argue with Aquinas or Ockham (which would be amazingly presumptuous) but simply to show that the common claim of the atheists, yet again, is factually wrong. Religion does not abandon rationality. Theologians can be masters of reason, and they are fully aware of the subtle and difficult relationship between reason and faith.

But, equally obviously, religious people do not confine themselves to "pure" reason—and certainly not in the narrow way that scientists do. The best scientists do not take a narrow view because they are narrow minded, but because they see this as the necessary tactic to gain insights that are as robust as possible. Similarly, the serious prophets do not seek mystical experience because they want a thrill, but because they see such experience as the greatest exercise in prehension that can be achieved—the direct fusion of the human kind with the mind of God (or as some modern physicists might say, with the mind of the universe). Theologians do not advocate faith as an indulgence, an invitation to partake in fairy-tales, but as a means by which we can gain greater insight into the way the universe really is, and why it is. Faith itself is not reason, but it is perfectly reasonable to embrace it, or to seek to do so. Contrariwise, to dedicate oneself unswervingly to pure reason, and to the measurable observations and directly testable hypotheses on which the atheist-scientists insist, is not itself unequivocally reasonable—not, that is, if you are interested in truth, and in living as well as possible in this universe.

Finally, of course, religion happens because people want it to happen, and because (as has often been shown) societies with robust religions tend to be more stable, and societies that are more stable tend, by definition, to endure. Atheists again like to put these undeniable facts in the worst light. People at large are attracted to religion, they argue, because they like the warm glow it can bring. Religious societies can be more stable because religion is deployed as an agent of oppression, either by theocracies or by secular states that employ priests in effect as policemen, or indeed as Orwellian thought-police.

The bad things are true, in the sense that they happen. But they do not represent all the truth, and certainly not the most important aspects of it. We could be equally cynical about science itself, which after all is the sole source of atom bombs and serious chemical warfare. Scepticism is necessary but out-and-out cynicism is merely childish. Science does do bad things but at its best it opens our eyes to the wonders of the universe, as John Ray would have said—and indeed as Richard Dawkins says. Science is also the source of vaccines and hotlines via email to the world's great libraries and to a great deal more besides that make it possible for human beings to live life to the full.

Religion viewed through similarly unbiased eyes emerges as the personal journey that all of us might undertake to gain some insight into why we are here; deeper insight than we can gain through a life without contemplation. It is the journey, too, by which we might become better people, living better, less selfish lives. Religion can be used oppressively of course but it also spells out a morality that seems in essence to be unimprovable and which holds societies together not through fear of hellfire but through consensus. Aristotle opined that people are naturally good, and prefer to do good; and religion has shown, with a few deft statements, how goodness may be defined, and how it may be achieved.

Of course religion is worthwhile. Actually, of all pursuits, when taken in the round, it is the most worthwhile of all.

A great many people have helped me to shape my thoughts on science and religion, and I am grateful to all of them. But I am indebted most of all to my wife, Ruth West, who has listened to all my ramblings and never fails to suggest new and fruitful ways of looking at things.

As for the literature: I have been inspired in particular by John Hedley Brooke's Science and Religion*;* The Passion of the Western Mind *by Richard Tarnas; and Arthur Koestler's* The Lotus and the Robot. *I am also a great fan of F.C. Copleston.*

Binsey, 2006.

F. David Peat

David Peat *received his PhD from Liverpool University, carried out research at the National Research Council of Canada and was a consultant for the Science Council of Canada. After meeting the physicist David Bohm in 1971 they collaborated on a book,* Science, Order and Creativity *and were working on a second at the time of Bohm's death. In 1996 Peat moved to the hill top village of Pari in Tuscany where, together with Maureen Doolan, he opened the Pari Center for New Learning. Peat is the author of over twenty books, the most recent,* Pathways of Chance, *is published by Pari Publishing.*

"Parallel Lines" is an updated version of an essay previously published on the Metanexus Global Spiral, www.globalspiral.com

PARALLEL LINES: FARADAY AND PASTEUR

F. David Peat

In writing to Ada Lovelace (Lord Byron's daughter and the programmer of Charles Babbage's "analytical engine"—the ancestor of all computers) the physicist Michael Faraday wrote, "I do not think it at all necessary to tie the study of the natural sciences & religion together and in my intercourse with my fellow creatures that which is religious & that which is philosophical have ever been two distinct things."[1]

Faraday was one of the greatest scientists of his age. There is little need in this essay to reiterate the range of his studies. Except perhaps to add that his notion of lines of electrical and magnetic force led James Clerk Maxwell to the theory of electromagnetism. In turn, Einstein's attempt to reconcile Maxwell's theory with Newtonian physics provided the key to the theory of relativity.

Faraday came from a very poor family—his blacksmith father was often too ill to work. As a result the boy received only a very rudimentary education and at the age of 14 in 1805 became apprenticed to a bookseller and bookbinder, a position that offered him the opportunity to engage in a program of self education. In particular, he was fascinated by the article on electricity in the third edition of *Encyclopaedia Britannica* and made himself an electrostatic generator from old bottles, and began to create his own experiments.

A key moment in his life came when he was given a ticket to attended Sir Humphry Davy's chemcal lectures at the Royal Institution. Faraday sent Davy a bound copy of his detailed lecture notes and asked if he could work as Davy's assistant. The great chemist later claimed that Michael Faraday was his greatest discovery.

Faraday, and his remark concerning the distinction between science and religion, set me thinking about another highly practical scientist, Louis Pasteur, who once said "the more one attempts to approach God through science the further one distances oneself from Him . . . there is an insuperable gap between science and metaphysics. Experimental science is essentially positivistic and has nothing to do with the essence of things."[2]

Despite the apparent dismissiveness of their remarks, both men were deeply religious; faith illuminated their work and their love of truth lay at the origins of the quotations above. The intellectual rigor of their approach should therefore warn us of the danger of making connections between science and religion that are facile and superficial. To enter the laboratory with a handful of comforting metaphors does a disservice to both religion and science. Similarly it is unwise to use scientific discoveries and theories as ways of justifying religious belief.

Faraday belonged to a sect of Dissenters known as Sandemanians. The origins of Dissent lay in views of Newton and Locke, in that their members reject the mysteries of religion and reliance on external authority and sought to approach "the word of God" through reason alone. Sandemanians for their part believed that Christ's world was exclusively spiritual. Faith was a matter for the intellect. It was an acknowledgment of religious truths, rather than one involving an exercise of the will or a desire for redemption. Indeed, they believed that the more one turned inward in search of faith, or in the desire for an experience of conversion, the further one moved away from God. Sandemanians, in rejecting the established church, pointed out the spiritual dangers for those who were more interested in their experiences of Christ, rather than with Christ himself. In this spirit of rejection can also be found the seeds of the Dissenter's political radicalism and of their support of the French revolution.

For Sandemanians, Christ's domain was distinct from that of the material world, nevertheless His imprint could be found in physical law. This factor became a major influence in Faraday's science. Above all, Faraday believed that divinity should be reflected in the unity of nature. Hence his observations of electrical effects produced by a moving magnetic field led him to conclude that reciprocal magnetic effects should be produced by an alternating electrical current. So electricity and magnetism should be two aspects of one underlying unity. Indeed, Faraday went even further.

In seeking to discover a similar unity between light and magnetism he attempted to determine if magnetic fields would affect polarized light—this yielded only negative results. He even tried to discover a link between the force of gravity and those of electricity and magnetism by dropping an electrical coil through a gravitational field. In essence Faraday was seeking a totally unified theory of nature of the type later envisioned by Einstein and his successors.

Since, for Sandemanians, God's word was direct, so too the forces of nature should be direct. There should be no gaps in nature. Rather than accepting the Newtonian notion of "action at a distance", Faraday believed that empty space was not a mere absence, a "nothing," but a "something", some-thing that was capable of supporting electrical and magnetic forces. In this way the seeds were sown for the important concept of "fields of force", a notion that was later refined by Maxwell. And, just as Sandemanians sought to "enlighten" others, so too Faraday, in his Christmas lectures at the Royal Institution, sought to enlighten the general public about the nature of physical law.

While the impulse for Faraday's science came from a religious belief in the unity of nature, nevertheless the scientific confirmations he achieved could not, in turn, be used as a basis for the support of his religious beliefs. God's law was one thing, but the domain of Christ and His truth were quite another. Pasteur appears to have been making a parallel observation in arguing that religious truths have no business intruding into the observations and theories of science. But let me return to Pasteur's quotation above, about the insuperable gap between science and metaphysics. In fact I must admit to being guilty of giving only the first part of what Pasteur said, for he went on to add that "this does not mean that we are forbidden to think about metaphysics or to believe in the God of the evangelists." Within the world of science, Pasteur felt, one can see "the presence and the beginning of proofs of the existence of God". Likewise when we come to the notion of infinity within the world " the supernatural is at the center of all our hearts". Likewise a visitor once asked Pasteur if he ever prayed, without looking up from his microscope the great Pasteur replied, "I'm praying now"

Both men, in speaking of the gap between science and religion, are also talking about a certain sense of purity—the purity of religion and the purity of science. For a Sandemanian such as Faraday, the case was clear. Religion was a matter of faith, the faith of the intellect in a religious belief that owed nothing to any external authority, be it the Church, the University or the Laboratory. In other words, science could not provide us with faith. It could not offer a proof for the existence of God. On the other hand religion, for a man like Faraday, could be the guiding force of a scientific life.

Sandemanians sought to preserve their faith from what they saw as the contamination that can arise when one relies on external authority, or becomes too preoccupied by one's own subjective experiences and desires. Likewise, Pasteur desired to preserve the purity of the scientific quest from what could likewise be called the contamination of subjective belief and metaphysical speculation. After all, he said, science does not tell us about "the essence of things, the origin of the world or its destiny". But at what point does a belief in, for example, unity, harmony, order and the pursuit of truth turn into contamination of the scientific quest? Only, I would hazard, when a belief becomes an inflexible dogma that is not accessible to critical investigation so that it ends up compromising the very principles outlined above.

Take for example, the case that Pasteur has become an icon for creationists (try typing in "Louis Pasteur" and "Creationism" into an Internet search engine!). It had been a simple matter for the French scientist to disprove the notion of spontaneous generation of life—a theory that held some currency at the time. But it is an enormous and unscientific jump of logic to extrapolate from a rather elementary demonstration to the claim that this implies that all life must have been the result of the biblical creation story and could have evolved in no other way. Neither is it legitimate to draw on Pasteur's speculations to then make the claim that species cannot evolve.

It is exactly here, when science is invoked to support some particular belief system or political doctrine, that the danger lies. A young couple may go to the gym every day, yet no matter how hard they exercise or how strong is their faith, there is no reason to believe that the baby they eventually conceive will become an outstanding athlete. Orthodox genetic science dictates that acquired characteristics are not inherited.

However the Soviet Union under Stalin badly needed to increase its yield of wheat. While the USSR had rejected religion as being "the opiate of the masses" it did have faith in the constant development of human society through dialectical materialism. But if society can be perfected, why not nature? The government supported the experiments of I.V. Michurin on the production of hybrid fruit trees which, he claimed, arose through the inheritance of acquired characteristics. Then T.D. Lysenko developed Michurin's hypothesis into what was to become official Soviet genetics, so official, in fact that those opposing Lysenko's ideas were arrested or disappeared.

By allowing Stalin's version of the Marxist-Leninist belief system to contaminate science it compromised its ability to ask questions and pursue paths of research. Not only did this do enormous damage to Soviet biology but it also made it difficult for any other researcher, who wished to question orthodox Mendelian inheritance, to be taken seriously.

The post-war British educational system involved a process of selection and of streaming based on the results of IQ tests carried out on eleven-year-olds—the 11-plus examination. This policy was deeply influenced by the research of Sir Cyril Burt which appeared to show that IQ arose largely through inheritance and had little to do with family environment or education. In other words, it made sense that some children are destined at birth for the professions and should therefore be educated accordingly, while others could be given an education more appropriate to "the working class". Naturally all this fitted quite well into Britain's mandarin class system.

Burt's early work convinced him that intelligence is largely inherited, with education playing only a smaller, secondary role. Influenced by such a belief he sought to produce a definitive demonstration and published his statistical findings on the IQs of a number of twins who, through adoption, had been brought up in different families. These figures showed once and for all that genetics, rather than nurture, was by far the dominant factor.

It was only after Burt's death that scientists became suspicious of the large number of twins he had been able to obtain for his sample. Their conclusion was that Burt had deliberately falsified his data in order to support a theory in which he so deeply believed, and one which also supported a particular British attitude towards class and education. Indeed it resonates so well with words from the hymn "All Things Bright and Beautiful":

> The rich man in his castle,
> The poor man at his gate,
> God made them, high or lowly,
> And order'd their estate.

This issue of two parallel lines—one being the pursuit of truth in the world of nature, the other the pursuit of truth in the world of the spirit, brings to mind a distinction made by Karen Armstrong in her book *The Battle for God*. This deals with the origins and rise of fundamentalism within the three monotheistic religions. Fundamentalism, she argues, has several roots, including social, economic and political pressures, but its essence is a series of shared beliefs that religious writings should be taken in a literal form. Armstrong sees this symptom as a confusion between *mythos* and *logos*. A misunderstanding that did not exist before the Enlightenment and rise of science.

Mythos involves the truths inherent in the world's great myths, stories of origins and religious texts. Mythos gives a society its deepest source of meaning. It connects an individual to the group, to a spiritual center and, in many cases, to the natural world. It produces harmony and unity within a society.

Mythos may be telling us something profound about the human connection to the earth, or about the birth of human consciousness, but its truths are very different from those of logos. Logos is concerned with the world of observable and objective fact. Logos is the world of science and physical law. Logos is to be interpreted literally. Yet while Logos may tell us about the "how" of things, it does not enter into the "why". Logos may tell us about the physical nature of our bodies, from cell to atom to elementary particles but, to paraphrase Pasteur, it can never tell us about our essential essence. Logos may describe the dynamics of a rainforest, or the structuring of a society, but it can never inform us of our deeper response to the natural world or to our sense of meaning at being in and contributing to our society.

The distortions of fundamentalism arise when mythos is treated as logos. That is, when myths and religious writings are treated like scientific texts that require literal and factual interpretation. Such readings inevitably produce profound distortions, as well as diluting the significance of what is being read. Both logos and mythos are required in our lives and work but their distinction should always be acknowledged.

The philosopher Whitehead said something similar when writing about the prehensile nature of human mental perception. We see the world in two ways he termed presentational immediacy and causal efficacy. In one, the world is present to us as a whole and imbued with meaning. In the other, we perceive the inter-connections between things and the details of the world, yet without any immanent sense of meaning. Possibly the former could loosely be identified with a religious perception of the world, and the latter with the scientific.

Likewise when we leave the office or laboratory to enter the world of nature our eyes constantly scan the world for meaning. A slight movement at the edge of our field of vision causes our eyes to move and so bring that area into focus on the yellow spot of the retina. This yellow spot has the ability to make much greater discrimination of detail. Yet while we seek to gather more detailed information about a particular object, there are always cells away from the yellow spot that remain sensitive to movements in the environment which could indicate some predator in our vicinity.

Our interaction with the world both visually and mentally always involves both aspects—we need to understand both the detail of the world as well as its inner meaning. Both are important, yet both cannot be present simultaneously, rather we must be involved in a constant move between the two. Likewise we have the two strands identified by Pasteur and Faraday. Could they be said to be two ways of looking at the one world, or one way, that of truth, of looking at two worlds?

References

1. Frank A.J.L. James and Michael Faraday, *The Correspondence of Michael Faraday* (London: INSPEC Inc., 1999).

2. For Pasteur's views on religion and science, see René Dubois, *Pasteur and Modern Science* (New York: Doubleday, 1960).

James F. Kemp

Jim Kemp is *a practising attorney in Sonoma County, California. He attended the College of Marin and San Francisco State University before receiving his law degree from San Francisco Law School and in 1972 a Doctor of Jurisprudence. He has argued cases before the California Supreme Court, the California Courts of Appeal and the United States Court of Appeal for the Ninth Circuit. Many of his cases are published in the California Official Reports and serve as precedents in various fields of law.*

FROM STARDUST TO ANGEL DUST

Remember you are dust and unto dust you shall return

A whimsical walk round the corner from stardust to angel dust

JAMES F. KEMP

Introduction Phase

> Round the corner
> Into the box circle
>
> Burnt Norton

In the first of "Four Quartets", Eliot takes us into the rose garden, into the place of our beginning and our end. But he does so by leading us via a magical path, *round the corner*, through the magic place, *the box circle*, and then into the rose garden. In like manner, this paper takes a whimsical walk into contemporary culture via an imaginary parable in which angels preside over an exploration of the findings of contemporary science and what it is those findings tell us about the world we live in and our participation in it. Right off, that seems a bad row to hoe. Angels and science? The two do not mix very well.

In this new millennium, however, scientists and theologians show signs of a truce that might end the hostilities between their respective disciplines. Contemporary developments in the physical sciences have reached fundamental questions customarily left to philosophy and other disciplines including religion. Questions such as: What is space? What is time? Even: What does it mean to exist? Many scientists do not like this non-empirical interrogation and some see it as the end of science.[1] Others are comfortable with the interaction and some even see it as good for science. Carlo Rovelli believes that occasional confrontations with those fundamental questions are indispensable to science.

> Newton and the founders of modern science debated those questions with passion and without that interaction with those questions, they would not have succeeded. Likewise, Einstein, Bohr, Heisenberg, Dirac and their colleagues wrestled with those questions when they laid the foundations for contemporary science. It is only the purest of the mid and late 20th century who believe that those questions sully science.[2]

That is pretty heavy stuff so let's make it clear from the start that this paper assesses the situation from a layperson's perspective, from the perspective of a non-scientist/non-theologian. And that's important because the experts' efforts will be fruitless if they do not take root in the general public and, if the experts' findings are incomprehensible to the general public, those specialists will have toiled in vain. This is especially true in today's pluralistic community: ivory towers may have housed the experts in the past, but they are now uninhabitable. If the experts are dealing with real reality, we ordinary folks will understand.[3] But we'll need help. Perhaps angels will lend a hand. Still, we need a method, a procedure.

I. The Science Phase

Scientific Research Program

So how do we do that; how do we fashion a parable with both angels and science as players? Well to begin we need a method, a procedure to follow. Since science serves as today's means for comprehending nature[4] and, since angels are messengers, we could use angels as vehicles for a whimsical scientific research program following the model of Imre Lakatos.[5] Those programs start with a central thesis—Lakatos calls it the hard-core thesis. It unifies the program. A protective belt of auxiliary hypotheses surround this hard-core central proposal and those auxiliary hypotheses both define and support the core thesis. Relevant data test those auxiliary hypotheses. Potentially falsifying data work to make changes in the auxiliary hypotheses rather than the core thesis, leaving room for adjustments without altering the entire project. The auxiliary hypotheses can therefore change with new data but they must continue to define and support the hard core thesis.[6]

Hard Core Thesis

The hard core thesis is this: ANGELS ARE MATURED UNIVERSES.[7] Angels emerge from the evolutionary processes that track the beginnings and development of universes—or better yet Allan Guth's term: pocket universes—within creation. They are the assimilation of the material stuff of universes with each universe's free, intelligent and conscious agents, as each universe looks to the ultimate source of its existence and the meaning of and the reason for its existence. Just as we humans have emerged from stardust in that same evolutionary process—stardust formed and constitutes the material for our human existence—we humans serve as the dust from which angels emerge. We are dust from which angels emerge in the same sense that heavy atoms formed in the interior of stars were explosively born into the universe as stardust from which we humans emerged. Just as stardust evolved in such a manner that human creatures emerged, so we human creatures may evolve in such a manner that an angel will emerge. In that case, an angel is the experience and expression—the realization of the meaning—of the interactions between creatures at a lower level of organization but something much more than the mere sum of those creatures and their interactions. An angel is therefore a communion[8] not simply an entity specially created by God. An angel too emerges from the evolutionary processes that mark the function of creation.[9]

Auxiliary Hypotheses

That is a challenging thesis for sure and, right away, it demands at least three auxiliary hypotheses. First of all, since tradition tells us that there are countless angels, there has to be a sufficient number of universes to populate the angelic host. Secondly, the thesis needs diversity. No two angels are alike, so each pocket universe should differ from the others in some fundamental way. Thirdly, the thesis needs a process whereby those diverse universes can mature into angels. Contemporary science has done just that, it has covered all bases: in the first instance, with various theories or conjectures generally described as metaverse or multiverse[10] consisting of a virtually infinite number of pocket universes, and in the second, the cosmic Landscape which accounts for the huge number of different possible environments in which pocket universes[11] can exist and, in the third instant, cosmic evolution We will briefly consider each.

"Multiverse": The notion of many universes goes back to Hugh Everett's "Many Worlds Interpretation of Quantum Mechanics" that postulated that at each collapse of a probability wave function, the universe splits into as many universes needed to accommodate each probability. But then, that was back when some physicists still wondered about the meaning of quantum physics. Today, cosmology too is leading science to many worlds, to the multiverse or megaverse, or regions hidden from other regions, or pocket universes; scientists have many approaches to the issue and many names for these independent regions—part of the problem has to do with pluralizing the word "universe." We will start with Andrei Linde's eternal or chaotic inflation: during inflation, like opening a bottle of soda pop, bubbles form and each bubble is an isolated region that evolves independent of the other bubbles, i.e., a pocket universe. Also, the multiverse expanded faster than light speed so each Planck region evolved independent of the others because light—information—did not have enough time to travel from one region to another. We can therefore treat each of those regions as a pocket universe.

Next, black holes are the birth process of new pocket universes. So, pocket universes with black holes mother baby pocket universes. To change metaphors for the moment, the singularity at the center of black holes acts as a sump pump drawing the information that fell into the black hole's boundary, through the singularity, out into the abyss to form a pocket universe of its own. Lee Smolin introduces an evolutionary perspective into this picture.[12] He points out that if black holes are the birth process for new universes, then natural selection would favor universes with lots of black holes and since the initial conditions and laws and constants that favor black holes also favor conditions that support life, natural selection would favor life-supporting pocket universes. We should, therefore, expect to find many of these fecund pocket universes with conscious, intelligent life. This is not only an interesting hypothesis, but also a testable, falsifiable, hypothesis.

String theory and its mysterious progeny M-theory also contributes candidates for nascent angels, *branes*. Branes are multidimensional surfaces to which most particles and forces (but not gravity) cling, thus each, more or less, constitutes its own world. They are floating about in a higher dimensional spacetime called the bulk. According to Lisa Randall, there could be, or are, an infinite number of these branes floating about in this ocean of existence.[13] For angelistic evolutionary purposes, these branes serve the same purposes as pocket universes.

Well then, science assures us that existence consists of endless pocket universes, or their look-a-likes, and that auxiliary hypothesis supports our hard-core thesis: there are plenty of pocket universes to populate the infinite angelic host with an infinite number of angels. What about diversity? Are those endless pocket universes clones or even look-a-likes? If they were, that would not support our hardcore thesis, for angels are their own persons, uniquely different from each other. What about these pocket universes? If the laws of nature were absolutely fixed and inflexible, then each pocket universe would be redundant and each angel would be singing the same old song, "here comes another one just like the other one". Recent developments in physics and cosmology have turned on its head this hoped-for Holy Grail of a single unique theory that determines everything. Diversity reigns in what Leonard Susskind calls the cosmic Landscape.

Distinguishing Susskind's Landscape from all those multiverse theories offers a good approach to the vastness of creation these two interrelated concepts open to us. The megaverse deals with the huge number of pocket universes, or regions, or bubbles, or areas—all existing isolated from each other whereas the Landscape deals with the huge number of different environments that pocket universes can exist in. So, the Landscape comes first but it is not a time or a place nor does it consist in a huge variety of times and places. It exists as a huge variety of possible environments in which pocket universes can exist. Figuratively, it is a landscape of all conceivable mountains, valleys, rivers, canyons plains, etc., etc. Technically, the elevations represent vacuum energy levels and the relative stability of those energy levels. Those on mountain peaks are unstable as the slightest whiff can set them off in one direction or another. Those on slopes are already in the midst of rapid change. Only the vacuum energy levels resting in valleys are sufficiently stable to sustain a pocket universe and determine its various particles with various masses, charges, spins and all its laws and constraints. Currently, string theory, rather than providing *the* unique consistent theory of everything, has provided 100^{500} of these possible environments (theories?) and it's still counting. The failure of string theory to come up with some elegant mathematical principle that uniquely and inevitably determines the laws of nature once and for all, offers us 10^{500} or 10^{1500}—or some other outrageous, unthinkable number—solutions to its equations that describe different independent universes or background spaces or environments to house universes. This inestimable plurality of independent regions or space/times all have different constituents, laws, constants; they each have their own set of interactions.[14] The Landscape is an infinity of possibilities; pocket universes are an infinity of actualities. That infinite landscape and its infinite population promise an infinitely diverse host of angels.[15]

So far so good: we have a huge population of pocket universes that are as different as chalk and cheese. What about the third auxiliary hypothesis, cosmic evolution? Does the way matters have evolved up to now suggest that evolution promises great things in the future, perhaps angels? Yes it does.

When inflation came to an end, the energy that had been driving the inflationary expansion was converted into mass and, in the high temperature at the time—about 10^{-10} seconds after the Big Bang—this resulted in a gluon-quark soup where all four fundamental forces were unified and the quarks behaved as free particles. After about 10^{-4} seconds, the gluons trapped the quarks into triplets to form neutrons and protons. After about three minutes, the temperature had dropped sufficiently for the neutrons and the protons to combine and form nuclei, but things were happening so fast there was only sufficient energy and time available to form the lightest nuclei. Many of the protons remained free and eventually went on to form hydrogen. The neutrons and the remainder of the protons combined to form helium. So, at this early stage, only hydrogen and helium populated the universe. The heavier atoms had to wait for their own dramatic birth process.

Nothing significant happened for another 300,000 years because the energy of the photons remained high enough to prevent the electrons and nuclei from forming atoms. After 300,000 years, the photons had lost much of their energy and the electrons and nuclei were then free to combine into neutral atoms, and radiation and matter disentangled and the universe became transparent. Before then, it had been opaque, nothing could be seen in or through it.

Gravity was now the dominant force. Gravity gathered together the tiny primordial quantum fluctuations generated during inflation and they grew in size. The universe became increasingly lumpy and islands of relatively dense matter gradually developed throughout the universe. These islands were not precisely uniform and they fragmented into many separate mini-islands. The islands formed galaxies and the mini-islands formed stars.

The temperature of the matter in these mini-islands increased as they collapsed on their way to stardom. Their centers became so hot that hydrogen nuclei were able to fuse together to form helium. This conversion of hydrogen into helium released enough energy to prevent further collapse of the mini-islands and they became stars. The helium nuclei in the core of the stars then fused together and formed the carbon, nitrogen, oxygen and other heavy elements so essential for life on Earth. The massive stars finally exploded as supernovas blowing away their outer regions into empty space as the stardust of our metaphor. The stardust from one of those supernovas in turn coalesced and formed our Solar System about five billion years ago after which life and we conscious, free, intelligent creatures emerged from the process. Are angels next? Are we, the conscious interacting relationship of that stardust, the angel dust of our metaphor? Will we evolve into angels? Perhaps.

That then gives us three auxiliary hypotheses to support our hard core thesis: a host of universes, each universe diversely unique and a compatible evolutionary pattern. Since, however, this little "research program" drags science well beyond the fundamental questions Rovelli had in mind into the depths of religion, the program needs a few words on the relation between science and religion. How, if at all, have science and religion interacted throughout the history of Western Culture from which contemporary science emerged and attained a global status?

We will therefore consider a brief overview of the historical interaction between science and religion before tracking this more religious process leading to angels. It will start with the Greek and Biblical perceptions on the issue, then Christian perspectives, and then move into the foundations and development of modern classical science and will end the survey with contemporary perspectives arising from relativity, quantum physics and 20th-century cosmologies. The paper will then focus on religious issues. Love will emerge from this as the key player and the paper will propose kenosis as the fundamental basis of love, kenosis not altruism.[16] The paper will then attempt to demonstrate how its hard-core thesis just might serve as a synthesis for contemporary science and religion.

Science & Religion: a Quick Study

Religion played an upper hand in the Greek era particularly with the Pythagoreans who saw the universe as intelligible and necessary. One could, therefore, deduce its particulars from general principles without looking at the details of the natural order. This top-down approach to the natural order left science as a form of religious contemplation.

The Bible also teaches that the universe is intelligible, but it introduced the idea of a free creator and, therefore, a contingent universe that did not have to conform to some preexisting form. Since nature could take on most any form, one could not entirely deduce its particulars by merely contemplating its intelligibility. We also had to look at her structure and function in this world if we wanted to know it. Nature could have been different from what it is: we have to pay attention to it, as it exists in this world, if we want to know it. This opened the door for empirical science.

The early Christian era, however, did not enter through that door. The early church—the age of faith—focused attention on humanity's relationship with the divine and internalized that relationship. It dogmatically defined that relationship without looking at the particulars of nature. Earthly existence was just a sojourn on our journey to the heavenly kingdom so its particulars warranted little attention.

It wasn't until the Middle Ages—the age of faith and reason—that the scientific investigation of nature through observation and reason was seen as an important component in our knowledge and understanding of reality. According to Richard Tarnas "Medieval theologians came to realize that an empirically rational exploration of the physical world could disclose nature's inherent religious value".[17] Thomas Aquinas believed that an understanding of natural existence would fully connect the created world with God.

Unfortunately, that synthesis of faith and reason, of religion and science, barely touched the ground before the notion of a double-truth universe displaced it and led to the belief that there was one truth for religion and another truth for science and the two did not mix. William of Ockham taught that there was no possibility of moving from a rational apprehension of the facts of this world to conclusions about God or other religious matters. Francis Bacon agreed. He discarded religious contemplation and propounded a new salvation for humanity within the natural order through the natural sciences' pursuit of the truth of the natural order. But, we not only had to know nature, we had to possess her and control her and, if she would not freely reveal herself to us or obey us well then, according to Bacon, we had to put her on the rack.

> Through the careful observation of nature and the skillful devising of many and varied experiments, pursued in the context of organized cooperative research, the human mind could gradually elicit those laws and generalizations that would give man the understanding of nature necessary for its control. Such a science would bring man immeasurable benefits and reestablish that mastery over nature he had lost with fall of Adam.[18]

That led into the Modern era in which classical science became the exclusive access for human understanding of nature and it reigned supreme until the 20th century. As Xavier Zubiri observed, science was our exclusive hope for an understanding of nature (see note 4).

Modern science then shifted our perspective 180° from religious contemplation to empirical observation. We will therefore turn to the shifting foundations of this new secular science: First classical science, the Newtonian syntheses that followed Copernicus' dismantling of the Ptolemy/Dante centuries' old model of the universe, and then contemporary science which, some four hundred years later, undercut the foundations of classical science. We will not here consider how or why the dust disturbed by that contemporary disturbance has not yet settled. That requires its own paper.

Modern Classical Science: Secular Absolutes

Just as the early Christian era was a system of absolute dogma, and in some instances Christianity still is, classical science—science from Newton to quantum physics—is a system of absolutes.[19] The scientific absolutes include time, space, determinism, reductionism and realism. Space was the absolute background against which the absolute fundamental entities that alone comprise everything that exists, played out their absolutely determined existence. Time was the absolute and invariant measure of change. Time measured change in location, and change in the states of systems in accord with the same tick-tock duration throughout the whole universe forever. The laws of nature were absolute. Absolute realism defined the manner in which we humans know the reality of that existence: scientific theory precisely represents reality on a one-for-one relationship. All these absolutes affected things—everything—but were themselves unaffected.

Absolute space and absolute time serve as the background against which nature is structured and functions, in which nature exists. Space and time are universals independent of any and all externals. Everyone experiences space and time identically without regard to their relative positions or their relative motions. This absolute space and absolute time was and is always there, always the same, and things occupy this absolute space and move and change within it in tempo with the absolute and invariable tick of universal time. Classical science held space and time as axioms: self-evident and universally recognized constants that needed no demonstration or experimental verification and outside of which nothing existed nor could exist. This was very harsh on angels for angels do not occupy space nor do they wax and wane in time. Angels, if they exist, exist a-spatially and a-temporally so classical science had no time or place for angels.

Absolute reductionism and absolute determinism served as classical science's axioms that define and explain the structure and function of reality. The entire universe—all reality—was a great machine[20] constituted by the sum of its parts and nothing more. One could take the machine apart piece by piece right down to the fundamental, indivisible parts and that was all there was to it. Those irreducible particles, those little ball bearings, constituted the fundamental entities that fashioned all that exists; there was nothing else except the space they moved in, the time during which they moved and the forces that moved them. That was it.

Absolute determinism governs the operation and function of those parts and the machine itself. Consequently every state of the universe and all its parts, at all times and in all places, is preceded by another state that absolutely determines what follows. It is therefore, in principle, possible to determine and know with absolute certainty the state of the entire universe at any time and place in the past or at any time and place in the future once we measure and know the details of its current state. Absolute reductionism absolutely determines the structure of the universe and absolute determinism unconditionally rules its function.

This bodes badly for angels. Angels, if they exist, exist as complexities that constituted far more than the sum of their parts and they function freely. Moreover, if we can predicate any attributes to angels, novelty is high on the list but classical science's machine permits no novelty. Once we have unveiled the elusive elementary particle, absolute reductionism absolutely determines the constitution of everything that exists and, once we have determined their positions, masses and motions at any place and at any time, absolute determinism absolutely determines the past, the present and the future of existence. There is no time or room for novelty or for angels.[21]

Under this scheme of things, the arena (spacetime) was the fundamental thing, the game was secondary and the rules of the game preconditioned the entire course of the game. Try telling that to sand-loters who improvise arenas wherever and whenever they come together. Try telling that to players whose interactions dance, bump, dodge, twist, turn, throw and catch in accord with their own unpredictable and surprising innovations. No, the game is the thing; all else springs from the game.

How is it that science succeeded in selling tickets to its absolute arena of predetermined positions and movements? There are good and sufficient reasons for this strange turn of events and, although they have deeper antecedents, science succeeded in selling those tickets after Copernicus displaced earth from its central resting place and set it in motion. Chaos followed. In order to restore some sense of order, the world needed a new time and place for things to happen. Newton—building on Galileo, Kepler and others—did that brilliantly. His (their) example can help us adapt to the new disturbance that has dismantled classical science's arena and stadium clock.

Contemporary Science: Secular Uncertainty

Contemporary science calls all those absolutes into question and opens doors through which we see creativity and novelty in nature. Perhaps, after all, there just might be room for angels. Relativity questions the concepts of space and time, and quantum physics questions the notion of existence itself. For relativity, space and time, or spacetime, expands, contracts and warps depending on the relative motion of points of reference; spacetime is gravity, it is the dynamic gravitational field. Quantum physics discovered that probability, indeterminacy and uncertainty lie at the heart of existence. Although noting an ongoing debate about what quantum theory really means, Brian Greene sums it up nicely

> By 1927, therefore, classical innocence had been lost. Gone were the days of a clockwork universe whose individual constituents were set in motion at some moment in the past and obediently fulfilled their inescapable, uniquely determined destiny. According to quantum mechanics, the universe evolves according to a rigorous and precise mathematical formalism, but this framework determines only the probability that any particular future will happen—not which future actually ensues.[22]

Furthermore, the great successes of contemporary science have given rise to a sense of science's limits and a realization of its involvement in the problems plaguing contemporary society. Richard Tarnas observes that "the aggressive exploitation of the natural environment, the proliferation of mass destructive technology, and the close association of scientific research with the political, military and corporate establishments belie science's traditional self-image of detached purity." This opens the previously reliable scientific world to fundamental questions. Tarnas goes on

> That conceptual framework was evidently both creating and exacerbating problems for humanity on a global scale. Scientific knowledge was stupendously effective, but those effects suggested that much knowledge from a limited perspective could be a very dangerous thing.[23]

There is therefore a clarion call in contemporary culture—a globalizing culture—for religious, theological and philosophic perspectives from which to draw knowledge and some understanding of the nature of existence served up by contemporary science. It's probably angels blowing their horns. They want to help cultivate the ground for an integrated, harmonious perspective.

This need for a new point of view also challenges science to revisit fundamental questions and move beyond its worn out notion that at its foundational level nature consists of fundamental things moving about against a background of fixed and absolute space and time. The hard-core physical sciences face a new frontier and classical science's machine model of the universe cannot carry it beyond that frontier. That machine still serves us in our ordinary explorations of nature on her local paved surfaces, but it is not the all-terrain vehicle we once believed. It cannot negotiate the terrain beyond this new frontier.

So then, relativity and quantum physics have led science beyond classical science's rigid background of fixed and absolute space and time against which nature played out its role in invariant compliance with absolute determinism and absolute reductionism. A rather dull and unimaginative time and place with very little room or time for the transcendent. These contemporary discoveries have opened nature to exciting novelty, where it enjoys springing new and fresh harvests of existence into the universe.[24] It is quite playful, not at all stoic.

But, what about this new frontier? What might lie beyond it? Well, these two pillars of contemporary science—general relativity and quantum physics—although they for sure point out the proper direction, they will not play together on the same field.[25] While they do not contradict each other, they cannot apply to the same system at the same time and place. General relativity describes a smooth continuous flowing system of events whereas quantum physics describes a teaming system John Wheeler described as *quantum foam*. Quantum theory accounts for the discontinuous and uncertain nature of matter and energy at and below the atomic level, whereas general relativity accounts for a smoothly warping and bending time and space on cosmological levels. Consequently, each is incompatible with the other in their respective domains and science is struggling for a new theory that incorporates both. Its name is quantum gravity. A theory of quantum gravity is the holy grail of the contemporary physical sciences; Lee Smolin says that "The search for quantum gravity is a true quest".[26]

That same Lee Smolin describes that effort in his earlier book *Three Roads To Quantum Gravity* (New York: Basic Books, 2001) and in Chapter Four he presents one of the essential cornerstones to any theory that could do the job: Entities—objects—like people and cultures are not really things, they are processes—the smallest units of which are events—unfolding in time. They cannot be comprehended as static objects. This universe of events is a relational universe: networks of relationships between events constitute the universe. In this relational universe verbs reign, nouns serve. Smolin concludes the chapter:

> In the earlier chapters I argued that our world cannot be understood as a collection of independent entities living in a fixed, static background of space and time. Instead, it is a network of relationships the properties of every part of which are determined by its relationships to the other parts. . . . This means that the world is not made of stuff, but of processes by which things happen. Elementary particles are not static objects just sitting there, but processes carrying little bits of information between events at which they interact, giving rise to new processes.

Well then, if Smolin is on to something, interactions are fundamental and inter-actors are derivative. So actions may replace objects as the fundamental element of our universe. Events, not things, lie at the most fundamental level of physical existence and their interrelationships constitute the cosmos and our participation in it. We are process! Not "things in process", just process. Even space and time are manifestations of actions within those processes. As we shall see, this new terrain seems a lot more hospitable to angels than the world we traveled in the Newtonian machine. As Eliot tells us in "East Coker", it's more like hearing music than playing a game of billiards—and angels make beautiful music but poor pool players.

> . . . or music heard so deeply
> That it is not heard at all, but you are the music
> While the music lasts.

Any such reversal of priorities between entities and events would lead to a deeper comprehension of who and what we are and who and what angels are and who and what God is and how they all interact and interrelate. And that leads us to issues usually left to religion.

II. The Religion Phase

Although Smolin will not allow for the transcendent—he's the Dawkins of the physical sciences—this new approach in science[27] has significance for those religiously inclined. Why? Because in any coherent approach to an understanding of reality, the book of nature should tell us things about God just as the books of scripture do. And, our comprehension of nature tells us not only about God but also about our existence in creation. If, therefore, creation is a reenactment[28] of its Creator and if creation is made not of things but rather the interrelation of events—happenings—then perhaps we err if we think of God as a being, as an entity or any object at all. If creation consists of interactions between events in relational processes, and if the beauty of creation points to a revelation of God, we approach a clearer picture of God by thinking of God as verb rather than noun.

Verbalizing God and Creation

Scripture points in this direction. Scripture tells us that God *Is*. When Moses asked "who shall I say sent me?" the Word of God said, "tell them that *I am who am* sent you".[29] The Old Testament thus understands God as the verb "to be". Jesus too, in whom the true God is perfectly manifested in creation,[30] describes himself as "I am".[31] Well then perhaps God is unqualified verb; not a being, not something, or all things, or everything, or anything at all. God just *is*. If so, our traditional view of God as the substantive "being" would be enriched by a complementary perspective: God is pure and simple isness. God is no subject or object or any noun or pronoun at all. God is pure verb, pure *isness*; God completely contains and express what God is in the "Who God is". Consequently there is not what God is, there is only God is.

This "event" perspective would then distinguish creation and Creator by contrasting the verb "to be" and the verb "to exist." Only God *is*. So what about the reality that is not God? Well, it *exists*. Consequently *God is and all else exists*. All that is not God *exists*; only God *is*. Good enough, but how, if at all, does existence relate to God? It does so just as the roots of the verb suggest. "To exist" is to await from that before which one stands. In light of this, all that is not God emerges into existence as a tendency to reenact Isness and continues to exist by standing before Isness and receiving existence from Isness. In the verb-based relational universe, everything that is not God stands before God and reenacts the *Isness* that is God. It exists. The important point is that universes, human creatures, and even angels exist, but they exist only because they reenact the Isness that is God.[32]

But how accessible is this God, who is apart from anything that exists? Pure *Isness* lies way beyond human experience, mental or physical. A lead from Einstein's equivalency principle might help us out here. Recall that Einstein's quickly formulated special relativity did not account for gravity and was limited to reference frames in constant linear motion. Years later, Einstein was still stumped in his attempts to include gravity in a more general theory. Then he had what he described as his happiest thought: gravity and accelerated motion are equivalent. We cannot distinguish between their effects and, if we cannot distinguish between them, they are the same process. He therefore set aside the inscrutable notion of gravity as a force with unlimited instantaneous effects and focused his attention on accelerated motion, something we can more readily comprehend, and he formulated his general theory of relativity; the theory of gravity, the theory that equates spacetime and gravity and, since gravity and accelerated motion are equivalent, the theory that equates—unites—accelerated motion with spacetime.

The Equivalency of To Be and To Love

In similar manner, the equivalence principle can help us consider[33] God. The naked, pure and simple verb "to be" is way beyond our experience and therefore nigh impossible to think about. But, to some miserable extent, we experience love and we certainly give it a lot of consideration. If then "to love" and "to be" were equivalent at the most fundamental level, the event perspective of love—God as the act, the complete yet always dynamic action of love—would provide us with access to God at a far deeper level than the usual approach of considering love as merely an attribute of a being—an entity—we call god. Although we don't have the empirical data Einstein had to demonstrate the equivalency of gravity and accelerated motion, the biblical injunction that God is love allows us to use the principle with respect to "to love" and "to be" and to comprehend the two verbs as equivalent. In that case, if we move from an entity perspective to an event perspective, we enhance our vision of God by viewing God as the pure verb "to love".

Simone Weil looked to God from this perspective. She concludes her essay, "The Love of God and Affliction", by pondering the question, "Why did God create? It seems so obvious that God is greater than God and the creation together." She sees light beyond the enigma by looking toward God from an event perspective:

> At least, it seems obvious so long as one thinks of God as Being. But that is not how one ought to think of God. So as soon as one thinks of God as Love one senses that marvel of love by which the Father and the Son are united both in the eternal unity of the one God and also across the separating distance of space and time . . . God is love, and nature is . . . a mirror of love.

So many things can be said about love. It is inexhaustible. But this contemporary era is coming round to seeing kenosis as love's period point. Not altruism, *kenosis*. Simone Weil leads the way on this road. According to her, God created through love, for love and created only love. The inseparable divine act of creation and salvation is kenotic love. Creation does not expand or extend God, nor does God command the existence of creation. Rather God withdraws from God, and into that opening "nothing" emerges from which existence springs forth and that which is not God flowers into existence.[36]

This is wonderful stuff, but it is a bit negative. Is it possible to access kenotic love from a positive point of view? Yes it is, and this is crucial. A positive approach sees kenosis as opening: opening self to the other. It's two imperatives then are *attention and consent*: attention as you must first know that the other exists and consent because you can love only what exists—or is—for otherwise you live in an imaginary world or, worse, you would have existence exist as you would have it exist and therein lies the roots of power and command.

From this kenotic point of view, God lets go of exclusive *Isness* by opening God to the existence of that which is not God, and God sustains creation in existence by consenting to and accepting the existence of that which is not God. As Weil puts it, God *abandons* existence, for if God did not abandon it, it would cease to exist. Creation then is not an exercise of power and might,[37] God does not command existence to exist. A god of power and might does not command the existence of creation. Rather, the true

God, the God revealed in the life of Jesus of Nazareth, exercises consent not command. Desire[38] for the other flows from the fountainhead of pure love and that desire engenders the existence of the other and then consents to and accepts its existence as it exists. We imitate the act of creation through a process of attention and consent.

The notion of a god of power and might has afflicted Christianity ever since Constantine's blasphemy when he pulled the Cross of Jesus out of its grounding in the earth and lifted it to heaven as a sword. This experience with Roman culture has soiled Christianity with a stain that all the blood of its history has been unable to wash out. Power and might, the power of command, must go. Seventeen centuries is enough. An event perspective offers us a view beyond power, might, command, force, violence and all their accompaniments—the list goes on and on; it would only end with total destruction, the undoing of creation.[39] Love, without exception, has nothing at all to do with power or powerlessness or anything else within that category. There is then no "power" of love. There is the energy of love, but there is no power of love.[40] In like manner, love has nothing at all to do with command; consent is love, command is rape.

This process of attention and consent dismantles in us that which expresses power and exercises command; that which would recreate the cosmos in our image and likeness, as we would have it, (*stones, be thou bread* [41]) as we would love it. Kenotic love seen from an event perspective will substitute attention and consent for power and command, and thereby open creation to angelistic evolution.

But then this approach leads to all sorts of problems. It asks us to accept things as they are, even if they are not "right". Does this mean we may not go about setting things right? What about the miserable suffering that afflicts so many of us? Must we consent to and accept it? For sure, we cannot love that ugly affliction, but perhaps if we embrace those it strikes down rather than striking down those we believe cause the affliction, we might strike a match to illuminate the heart of affliction. If we love the afflicted more than we hate the oppressor, the energy of love might set things right. The energy of love will warm those frozen isolated events and let them flow. That is the true essence of faith. Faith is not some sort of submission to creeds, doctrines or dogmas; *faith is trust in the efficacious action of love*. Jesus demonstrated that clear enough.

The Energy of Love

When the interrelation of the events and processes in creation reenact the action of love, the energy of love will transform existence just as radically as a drop in temperature transformed the universe from a soup of quarks and electrons into a system of interacting atoms. So long as those ugly self-centered events and processes are frozen and isolated into virtually static entities, love fails. They must interact; they must relate. Simone Weil tells us that existence in this world has meaning only in terms of relations:

> . . . the part of the soul that has seen God must transform every relation with a created being or thing into a relation between that being and God. Every relation between two or several created things—whether thinking beings or matter—is one of God's thoughts. We ought to desire a revelation of the thought of God corresponding to each relation with our fellow humans or with the material objects with which we are involved. . . . The end is to conceive each of them, specifically, as a thought of God.[42]

So then, as already mentioned, we can imagine such transformations with help from science's notion of *phase transitions*; like water transforming into ice or steam at critical temperatures. Our universe underwent phase transitions when the temperature dropped to critical points at which forces of nature separated and radically transformed the makeup of the universe. It is therefore, not entirely unreasonable to anticipate future phase transitions. Future phase transformations occur when temperature rises to a critical point and all the forces of nature unite. Only this time *the* fundamental force/energy accomplishes this rise in temperature. That energy/force is love: the love that is the isness of God reenacted in creation as existence.

This fits so nicely with an important work on Love by Pitirim A. Sorokin.[43] Listen to what Sorokin has to say about *ontological love*:

> Love is the universal creative force . . . [It] tends to make the whole universe one harmonious cosmos . . . . (p.8) . . . . Such a love cosmos is now only at its beginning in the empirical world of man, (sic) at about the same stage as is reason in the animal world (10-11)

The energy of love is the key concept in his book but Sorokin focuses on altruistic love whereas kenotic love may be a step further toward a better illumination of the subject. In Chapter Three, Sorokin looks at the energy of love in the same way we look at energy in the physical aspects of existence. He considers how we generate, store and distribute the energy of love. In the remainder of the book, he considers how the interactions of that energy/force transform all of existence.

The important point for our purpose is that we in our universe have not yet allowed this embryonic love to express itself in interactions within creation. As a result, this virtual absence of Love—the absence of God—has reduced those interactions to isolated frozen events. When, however, love melts those frozen events, they will release the enormous energy of love—like matter into energy—and we will find ourselves in a chain reaction exponentially releasing unimaginable energy. That is terribly important for the phase transition we will consider in a moment. But first, we need some understanding of the nature of love as we experience it in this world.

The Movement of Love: Separation and Unity

So we reenact love through a process of attention and consent rather than power and command, but how does it function, how does it work in a relational universe? Simone Weil tells us that love has two movements: separation and union.[44] Through this paradoxical dynamic of love, Weil offers insights into an action-based reality.

Although with God *inter se*, these separation and union movements of love are not distinguishable categories. From the creature's perspective, we see both perfectly realized in the internal dynamic that is God: the dynamic that is the perfection of Love. The infinite oneness within God generates separate Persons—and, in this sense, a person is a principle of action—who perfectly realizes and perfectly expresses the perfection of that oneness by uniting in infinite union, into identity. This dynamic serves as the foundation for our concept of Trinity.[45] The *inter se* action of God—Isness, perfect Love—constitutes an indivisible, non-durational dynamic in which oneness separates into persons who unite in perfect union, identity.

When we focus on the *ad extra* action of God, creation is the perfect expression of the separation movement of Love. But, what about the union movement of love[46] here below in creation? How does creation reenact the union movement of love?

Angels just might do that for us or, rather, the angelistic may be the mode through which we will approximate that perfection of unity as we move through ever more complex systems and an angel emerges from the evolutionary processes of this universe we inhabit; evolutionary processes in which we participate and from which we emerged.

III. The Angelistic Phase

So then, what and who are angels? What are these matured universes? What do they do and what is their purpose? Angels are the realization and the expression of the union aspect of love in creation. The separation aspect is realized and expressed *ad extra* by creation—the generation of that which is not God and its separation from God. So, creation is the realization and expression of the separation aspect of love. Created love therefore starts with separation and the union aspect begins and develops with the evolutionary processes within each universe and reaches its full realization and expression in an angel. Angels are the ultimate realization and expression of the union aspect of created love. As the separate and diverse interactions of created reality—the interrelationship of events—evolve into ever more and more complex relationships, those processes become more and more united, all the while retaining their diversity and ultimately attain the full realization and expression of their diverse unity—or unitary diversity—in angels. Angels continue the created act of love by uniting the diverse separation of created love in the essence that constitutes their existence as the ever more perfect reenactment of the infinitely perfect act of Love that is God. The whole of the universe, including every human person, unites as the person and the nature of an angel reenacting the *Isness* that is God in its *existence*. We humans are, therefore, brought together as one along with the rest of nature. And this coming together in communion realizes the unification movement of love in creation.

This, however, raises an appropriate concern: Do we lose our uniqueness in that unity? Does that unity dissolve diversity? That would not be a satisfactory state of existence. But again the answer is no. Instead, the process brings us to an exciting contradiction: unitary diversity or diverse unity. This is much the same contradictory diverse unity some express in their concept of Trinity. According to Simone Weil, contradiction is a sure sign of reality. She tells us, "The contradictions which the mind is brought up against form the only realities, the only means of judging what is real. There is no contradiction in what is imaginary."[47]

At the human level, compassion is the key to the process by which we participate in the union aspect of love and kenotic consent fuels compassion. This process, therefore, correlates nicely with Murphy and Ellis' hard-core thesis in *The Moral Nature of the Universe*: Kenosis is the fundamental law of the universe.

True compassion requires our kenotic consent to the existence of the other. This includes our consent to the reality of affliction generally and, like the Good Samaritan, our consent to the existence of the afflicted, as he or she exists. That consent is kenotic because it involves a letting go of self; a projection of that self into the afflicted. That projection, in turn, brings us the surprising news that *self* is an illusion. Self is a frozen event, a bit of opacity that stands between God and God's creation and when we consent to the afflicted we erase that opacity making it possible for God—Love—to enter into creation and love the afflicted, as the afflicted exits *through us*. It, like Eucharist, is true sacrament. So, it is not our *self* that enters into the afflicted, it is God who enters into her or him and that is made possible by our kenotic consent to the existence of the afflicted as the afflicted exist. Although our *self* does not enter into the afflicted, the passage of God's love through us into the afflicted unites us with the afflicted one. That one-by-one union will eventually unite all human persons and that ultimate union constitutes an angel of diverse unity, unitary diversity.

Does this angelistic union, or communion, bring our existence to some sort of end point? Does it point to the same sort of boredom that Newton's machine pointed toward the end of the 19th century? Not at all. Angels are never boring. Only imaginary frauds bore us. The unity that constitutes angels does not bring us to a static end of the road. Rather, it opens to us many new diverse paths that endlessly lead to new understandings and new experiences of the inexhaustible Isness that is God, Love. Jesus tells us that eternity is to know God.[48] To know God is impossible, but to experience deeper and deeper understanding of God is an endless joy that exponentially expands at each new level. Since the existence of creation is a reenactment of the Isness that is God, the whole process begins with our understanding of nature and our experience of its beauty. As Simone Weil puts it: Beauty is the explosion detonated when our sensibilities come into contact with the order of the world. This process will also include our experience of each universe in the eternal infinity of universes eternally generated by the Love that is God.

So the evolution of existence does not stop with angels. Just as the explosion of one generation of stars after another generated the stuff—the dust—from which we humans emerged, so too each angel, each matured universe, will unite into mega-angels—archangels?—and they too will unite in an eternal dynamic unendingly coalescing creation in ever more complex unity all the while coming infinitely closer and closer to the Face of God; eternally closer, but always infinitely distant.

Gregory of Nyssa described this evolution of the universe with his doctrine of *epektasis*, perpetual growth:

> The creation attains excellence by partaking in something better than itself; and further, not only had a beginning of its existence, but also is found to be constantly in a state of beginning to exist in excellence, by its continual advance in improvement, since it never halts at what it has reached, but all that it has acquired becomes by participation a beginning of its ascent to something still greater, and it never ceases, . . . *Contra Eunomium 3.6.*[49]

Gregory believed that evolution is the action of creation, consequently, an end of evolution would be the end of creation.

In keeping with this insight and with Smolin's verb-based universe, Diogenes Allen sees perfection as a process not a state:

> The Christian understanding of contemplation has to be freed from Platonic and Aristotelian assumptions that the goal of life is the contemplation of a static and never-changing reality . . . the question is this: do our lives . . . attain perfection when they cease to change. [The answer is a resounding NO] . . . perfection means a neverending increase in goodness and excellence, not stasis . . . we never cease to grow in excellence and goodness . . . Gregory . . . rejoices in growth and understands the goal of the Christian life as one of neverending growth and change, which is the meaning of perfection . . . Gregory of Nyssa thus enriches our understanding of neverending joy, since contemplation is not merely looking at God but the continuing development and growth of our own person as we interact with the source of all that is good and desirable.[50]

We can't close without acknowledging death. What about death? Death is the transition process leading to the angelistic phase of existence. This paper will not consider that transition in any detail or compare it with earlier phase transitions our universe experienced. It will settle on references to death from the bible and its favorite poet, references that equate death with birth just as Gregory did. Eliot begins "East Coker" with *In my beginning is my end* and ends it with *In my end is my beginning*. These lines and this paper's scheme of things illuminate the New Testament's vague notion of a new birth: any end is always a new beginning and those new beginnings are the eternal process of existence.

Conclusion Phase

This paper started in the first "Quartet" with Eliot's "walk round the corner . . . into the box circle". At the end of the last "Quartet", "East Coker", Eliot points to the purpose of that walk. That purpose is "a further union, a deeper communion". With that line, Eliot references lines from Dante, lines which Eliot described as the highest poetry has ever achieved and the highest it will ever achieve; lines which graced Western Culture with its emblematic Rose

I saw within its depths how it enfolds
All things in a single volume
Bound by love of which the universe
Is the scattered leaves.

Paradiso, Canto XXXIII, 85

Notes

1. Horgan, John, *The End of Science*, (New York: Broadway Books, 1996).
2. Carlo Rovelli *Quantum spacetime: What Do We Know*, p. 102 in Callender, Craig & Huggett, Nick (eds.), *Physics Meets Philosophy at the Planck Scale* (Cambridge: Cambridge University Press, 2001).
3. For example, Lee Smolin expects that quantum gravity will be taught to high school students in this century. *Three Roads To Quantum Gravity*, (New York: Basic

Books, 200?), p. 211. Quantum gravity is the most esoteric problem currently engaging the physical sciences.

4. Here's what a contemporary philosopher said on the subject, "Aristotle rigorously distinguished science from metaphysics thus proposing the non-univocal nature of realty. The modern era, since about 1700, lost that wisdom. Since then, we are so persuaded that reality is discovered to us only by science that we have made science a constitutive approximation to reality. This is the result of a profound conviction that in science we are served the only portion of realty which is accessible to us with certainty." Zubiri, Xavier, *Nature, History, God* (Thomas B. Fowler, Jr., trans.) (Lanham, New York: University Press of America, 1981), p. 71.

5. Imre Lakatos is a philosopher of science who synthesized the apparently conflicting ideas of Karl Popper and Thomas Kuhn and demonstrated that all sciences operate from a common "methodology of scientific research programs". In *On the Moral Nature of the Universe: Theology, Cosmology, and Ethics*, (Fortress Press, 1996), Nancey Murphy and George F. R. Ellis claim that both theology and ethics can be dealt with as sciences because they also involve this same sort of research programs.

6. David Bohm and David Peat describe the same process by distinguishing paradigm/theory and fragmentation/specialization and by debunking the notion of *absolute* truth. Bohm, David and Peat, F. David, *Science, Order & Creativity* (New York: Bantam Books, 1987), Ch. 1.

7. This is nothing new. In the 15th Century, Bruno looked to the heavens and saw a virtually infinite number of worlds out there each with its own soul. He was burned at the stake because authorities saw his vision as questioning orthodox understandings of the Incarnation and other religious dogma. They did not, as has been reported, execute him because of his science. Also I remember reading a story when I was quite young about some one who found a very special diamond and, rather than using or selling it as an ornament, he polished it into a lens through which he saw atoms as planetary systems with people on some of the planets. He fell in love with a beautiful young girl on one of the planets—he saw things with that much clarity, it was a wonderful diamond—and he was consumed by yearning for the inaccessible.

8. In its attempts to suppress religious pluralism, the Vatican has soiled this word by applying it to those who blindly submit to the dogmatic statements it makes on religious matters—and often on other matters properly left to science, philosophy and other disciplines. I shall nonetheless continue to use it in its traditional sense so full of beautiful meaning and symbolism.

9. Meister Eckhart sensed this angel interrelationship back in the 14th century: "Go away and sink deeply into yourself until you understand the angel and give yourself

up to that with all your being and realize that you consist of nothing else than what you find in that angel. Then you will realize that you are one with the angels. And when you give yourself to this realization with all your being, then it will dawn on you that you are all angel and with all the angels." Quoted by Matthew Fox in *Passion for Creation: The Earth-Honoring Spirituality of Meister Eckhart,* (Rochester, VT: Inner Traditions), p. 94.

10. See for example, Professor James E. Lidsey's account in *The Bigger Bang* (Cambridge: Cambridge University Press, 2000). Apparently, this proliferation of universes is a variation on the Anthropic Principle: Since life as we experience it is possible only within a very narrow range of constraints, is it pure luck that we exist or did some benevolence design our universe for us? Neither! It's probable because you have a large number of universes—certain, if you have an infinite number.

11. See *The Cosmic Landscape: String Theory and the Illusion of Intelligent Design,* Leonard Susskind, Back Bay Books, 2005. Susskind explains, "The *Landscape* is the term that I coined to describe the entire extent of these theoretical environments. The Landscape is the space of possibilities—a schematic representation of all the possible environments permitted by theory." . . . Each pocket [universe] has its own "weather": its own list of elementary particles, forces, and constants of physics." Susskind credits this landscape diversity with his conversion to the Anthropic Principle. p. 14.

12. *The Life of the Cosmos* (New York: Oxford University Press, 1997).

13. *Warped Passages* (New York: Harper Perennial, 2005) especially Chapter 3.

14. Smolin, Lee *The Trouble with Physics* (New York: Houghton Mifflin, 2006), Chapter 10.

15. It may be a bit anti-Copernican to classify environments as lethal. Susskind's grossly lethal pocket universes just may harbor some form of something akin to what we call consciousness and intelligence and we and they might interact angelistically. Mere naked existence is a level of relationality.

16. Early on I thought of altruism as a flashing red light: *caution, slow down, pause.* Now I am convinced that it is a glaring red light: *stop*; it is always necrophagous. It always feeds off the dead flesh of the afflicted to nourish its own sense of self worth.

17. Richard Tarnas, *The Passion of the Western Mind* (New York: Harmony Books, 1991), p 177.

18. ibid., pp. 272-273.

19. I look at the Promethean nature of Newton's appropriation of absolutes from the divine into the profane—his immanentization of transcendent absolutes—in a chapter on modeling the universe in a work in progress.

20. The contemporaneous industrial revolution played a role in fashioning this picture.
21. All of this is too negative. It's too harsh on Newton (not absolutes). Newton's synthesis of the chaotic motion left behind when Copernicus displaced earth from the center and put it in motion is one of the greatest achievements of the human mind. He couldn't have done it without absolutes. He saw Galileo's relativity as a problem the *Principia* solved. Newton wrote a margin note near the end of the *Principia* promising to deduce absolute motion from relative motion commenting "For it was to this end that I composed it [the *Principia*]". (See *Physics Meets Philosophy at the Planck Scale*, Craig Callender & Nick Huggett (eds.), (Cambridge: Cambridge University Press, 2001), Chapter 9, "On General Covariance and Best Matching", Julian B. Barbour, p. 200.) According to Barbour, Newton attacked Descartes' relativity not Galileo's. Descartes saw constant linear (inertial) motion as relative but assigned each moving object an absolute referent: its contiguous objects and thereby put earth at rest relative to the vortex that carried it around the Sun. (*The End of Time*, Oxford, 1999, pp. 61-2.) False but necessary ideas fill the history of the physical sciences. For examples: Newton knew it was crazy to believe a force (gravity) could operate instantaneously over huge distances. And what physicist real believes in dimensionless "point particles"?
22. Greene, Brian, *The Elegant Universe*, (New York: Vintage Books, 1999), p. 107.
23. Tarnas, *The Passion of the Western Mind*, op. cit., pp. 364-365.
24. Evolution opened the life sciences to this same excitement but that's for another time and place; for now, it's the physical sciences.
25. Science successfully united SPECIAL relative and quantum physics in relativistic quantum field theories that account for the interaction of three of the four forces of nature, electromagnetism, the weak force and the strong force. But, gravity and quantum physics just wouldn't mix. GENERAL relativity is the theory of gravity.
26. *The Trouble with Physics*, (New York: Houghton Mifflin, 2006), p. 80.
27. Process theology probably has a similar vision but it builds from rational discourse whereas Smolin builds from empirical data thereby opening religious awareness to contributions from contemporary science. Furthermore, for process thought, the divine has two natures: the primordial and the consequent. It sees a paradox in an immutable, all-knowing and all-powerful God who creates a contingent reality. It would relieve the tension of the paradox with these two natures: the unchanging primordial nature affects the world but is unaffected by it, whereas the consequent nature changes with the world as part of it becoming something? Or some event? Process thought succeeds by viewing God as verb rather than noun but it fails by importing duality into God. It fails by viewing this non-entity God as primordially

simple, on the one hand, and, on the other hand, a changing process participating in existence in such manner as to develop with nature into some sort of Teilhardian *omega* point. See John A. Jungerman, *World in Process*, State University of New York Press, 2000.

28. The bible speaks in terms of *image* "God created man in his image. In his image he created him." (Genesis 1:27) That imaging, however, is appropriate for an entity based reality but not an action based reality. *Reenactment* better fits a verb-based world.

29. Exodus 3:14.

30. Although the perfect manifestation of the true God in Jesus is universal, for religion, the pressing question today is whether that manifestation in Jesus is exclusive or are there others. See Roger Haight, *Jesus, Symbol of God* (Maryknoll, New York: Orbis Books, 1999).

31. John 13:19.

32. From a metaphysical perspective, Frederick Wilhelmsen traces the history of thinking about the act of existing apart from things that exist, but he limits existence to "things" thereby denying existence to the naked act of existence. See *The Paradoxical Structure of Existence* (Irving, TX: University of Dallas Press, 1970). From a scientific perspective, David Finkelstein strongly advocates an action based reality rather than an object based reality. See *Quantum Relativity: A Synthesis of the Ideas of Einstein and Heisenberg* (Berlin: Springer-Verlag, 1996).

33. Consider is a beautiful word. Its roots lie in the Latin for "with the stars".

34. *Simone Weil*, writings selected by Eric Springsted, *Modern Spiritual Masters Series* (New York: Orbis Books, 1998), p. 66.

35. See for example, *On the Moral Nature of the Universe: Theology, Cosmology, and Ethics*, Nancey Murphy and George F. R. Ellis, Fortress Press, 1996. In the last paragraph they affirm that kenosis is the fundamental law of the universe.

36. This same passive withdrawal permeates the Jewish mystical tradition on creation: "Luria pondered the question of beginnings. How did the process of emanation start? If Ein Sof pervaded all . . . , how was there room for anything other than God to come into being? . . . the first divine act was not emanation, but withdrawal. Ein Sof withdrew its presence "from itself to itself," withdrawing in all directions away from one point at the center of its infinity, . . . thereby creating a vacuum. This vacuum served as the site of creation." Matt, Daniel C. *The Essential Kabbalah* (San Francisco: Harper, 1994), pp. 14-15. See also *The Zohar*, Volume One, Translation and Commentary by Daniel C. Matt, Stanford University Press, 2004, p. 110.

37. What on earth have we Roman Catholics done to the beautiful Sanctus? Who substituted a god of power and might for the God of Hosts, and why? Saboath may

be rendered "army" but the heavenly hosts are not making war they are glorifying God. Michael did not vanquish Lucifer and drive him from the presence of God. Rather, God simply said to Lucifer "be it done to you according to your will." That's how Dante has it.

38. Desire has some wonderful nuances. Simone Weil tells us that desire is the presence of an absence and the dictionary tells us that desire comes from the Latin for "to await from the stars".

39. This is radical and not at all popular particularly in light of current notions of empowerment. Furthermore, it would shatter our cherished notion of *rights*. It reaches right down into the foundations of our political, economic and legal systems. They are all based on "rights" and "duties" And that is part of the problem. The notion of rights is inherently contentious and can be implemented only through the exercise of power and command. Weil was aware of this and, in her proposed constitution for the new France that would emerge after WWII, she proposed "needs" and their correlative "obligations" as the appropriate foundation for society. (*The Need For Roots*, Routledge, London, 1995.)

40. Actually, energy is a rather attractive category for love. In the physical sciences, energy and matter are manifestations of the same underlying reality. We may thus see matter as frozen energy, a frozen event. The distance that separates God from creation cools the energy of love into frozen isolated things. Here below we can warm them through love and, when we do, they melt into moving processes.

41. Matthew 4:3.

42. Springsted, Eric; *Modern Spiritual Masters Series* (New York: Orbis Books, 1998), pp. 89-90.

43. *The Ways and Power of Love*, Timeless Classic paperback (Philadelphia: Templeton Foundation Press, 2002), originally published 1954 by Beacon Press.

44. Weil develops this in her amazing essay, "The Love of God and Affliction". It can be found, with other significant writings concerning love in *Simone Weil*, writings selected by Eric Springsted, *Modern Spiritual Masters Series* (New York: Orbis Books, 1998), pp.41-71; see particularly p. 48.

45. But it is well to keep in mind that trinity is a human construct that can only reflect the divine dynamic if we properly relate it to the interrelation and interaction between God and creation. Trinity does not define and limit the internal structure of God. Rather Trinity expresses the interrelation and interaction between God and creature from the point of view of creature. Roger Haight, *Jesus, Symbol of God*, (Maryknoll, NY: Orbis Books, 1999), pp. 471-473.

46. Did God choose to create and does God choose to sustain creation in existence? Well, yes God did choose to call nothing into existence and God does choose to sustain it in existence because God is free to do so or not to do so. God is free

because freedom is embedded in the essence of love to such an extent that there is no love without freedom and, since God is Love, God is free. This means, at least in human terms, that God has a choice; God may create or refuse to create. Still, "to create" or "to sustain creation" isn't simply a matter of deciding whether to do something or not to do it. They also are inherent constitutive properties of love. This gift of existence is therefore a paradox. It is a free gift but it is also an unavoidable effect of love, an imperative of love.

47. Weil, Simone, *The Notebooks* (Arthur Wills trans.), (London: Routledge & Kegan Paul, 1952), Vol. 1, p. 329.

48. John 17:13.

49. Quoted in *Gregory of Nyssa's Doctrine of Theistic Evolution*, by Daniel F. Stramara. CTNS Bulletin, Vol. 22, Number 2, Spring 2002, p.13. Stramara calls this process *epektasis*; Roger Haight calls it a*pokatastasis* (*Jesus, Symbol of God*, op. cit., p. 393).

50. Allen, Diogenes, *Spiritual Theology* (Cambridge/Boston, MA: Crowley Publications, 1997), p. 32.

Society

John Scales Avery

John Scales Avery *is associate professor of quantum chemistry at the University of Copenhagen. He has been active in the Peace Movement and associated with the Pugwash Conferences on Science and World Affairs. In 1995 this group was awarded the Nobel Peace Prize. His books include* Information Theory and Evolution *and* Space Age Science and Stone Age Politics.

AGAINST THE INSTITUTION OF WAR

JOHN SCALES AVERY

In the first decade of the new millennium, our scientific and technological civilization seems to be entering a period of crisis. Today, for the first time in history, science has given to humans the possibility of a life of comfort, free from hunger and cold, and free from the constant threat of infectious disease. At the same time, science has given us the power to destroy civilization through thermonuclear war, as well as the power to make our planet uninhabitable through pollution and overpopulation. The question of which of these alternatives we choose is a matter of life or death to ourselves and our children.

Science and technology have shown themselves to be double-edged—capable of doing great good or of producing great harm, depending on the way in which we use the enormous power over nature, which science has given to us. For this reason, ethical thought is needed now more than ever before. The wisdom of the world's religions—the traditional wisdom of humankind—can help us as we try to insure that our overwhelming material progress will be beneficial rather than disastrous.

The crisis of civilization, which we face today, has been produced by the rapidity with which science and technology have developed. Our institutions and ideas adjust too slowly to the change. The great challenge which history has given to our generation is the task of building new international political structures, which will be in harmony with modern technology. At the same time, we must develop a new global ethic, which will replace our narrow loyalties by loyalty to humanity as a whole.

In the long run, because of the enormously destructive weapons, which have been produced through the misuse of science, the survival of civilization can only be insured if we are able to abolish the institution of war.

While in earlier epochs it may have been possible to confine the effects of war mainly to combatants, in the 20th century the victims of war have increasingly been civilians, and especially children. For example, according to Quincy Wright's statistics, the First and Second World Wars together cost the lives of 26 million soldiers, but the toll in civilian lives was much larger—64 million.

Since the Second World War, despite the best efforts of the UN, there have been over 150 armed conflicts; and, if civil wars are included, there are on any given day an average of 12 wars somewhere in the world. In the recent conflicts in Indo-China, the proportion of civilian victims was between 80 percent and 90 percent, while in the Lebanese civil war some sources state that the proportion of civilian casualties was as high as 97 percent.

Civilian casualties often occur through malnutrition and disease, which would be preventable in normal circumstances. Because of the social disruption caused by war, normal supplies of food, safe water and medicine are interrupted, so that populations become vulnerable to famine and epidemics. In the event of a catastrophic nuclear war, starvation and disease would add greatly to the loss of life caused by the direct effects of nuclear weapons.

The indirect effects of war are also enormous. Globally, preparations for war interfere seriously with the use of tax money for constructive and peaceful purposes. Today, despite the end of the Cold War, the world spends roughly a trillion (i.e. a million million) US dollars each year on armaments. This enormous flood of money, which is almost too large to imagine, could have been used instead for urgently needed public health measures.

The World Health Organization lacks funds to carry through an anti-malarial program on as large a scale as would be desirable, but the entire program could be financed for less than the world spends on armaments in a single day. Five hours of world arms spending is equivalent to the total cost of the 20-year WHO campaign, which resulted in the eradication of smallpox. For every 100,000 people in the world, there are 556 soldiers, but only 85 doctors. Every soldier costs an average of 20,000 US dollars per year, while the average spent per year on education is only 380 US dollars per school-aged child. With a diversion of funds consumed by three weeks of military spending, the world could create a sanitary water supply for all its people, thus eliminating the cause of almost half of all human illness.

A new and drug-resistant form of tuberculosis has recently become widespread, and is increasing rapidly in the former Soviet Union. In order to combat this new form of tuberculosis and prevent its spread to Western Europe, WHO needs 450 million US dollars, an amount equivalent to four hours of world arms spending. By using this money to combat tuberculosis in the former Soviet Union, WHO would be making a far greater contribution to global peace and stability than is made by spending the money on armaments.

Today's world is one in which roughly ten million children die each year from diseases related to poverty. Besides this enormous waste of young lives through malnutrition and preventable disease, there is a huge waste of opportunities through inadequate education. The rate of illiteracy in the 25 least-developed countries is 80 percent, and the total number of illiterates in the world is estimated to be 800 million. Meanwhile every 60 seconds the world spends roughly 2 million US dollars on armaments.

It is plain that if the almost unbelievable sums now wasted on armaments were used constructively, most of the pressing problems now facing humanity could be solved, but today the world spends more than 20 times as much per year on weapons as it does on development.

Because the world spends a thousand billion dollars each year on armaments, it follows that very many people make their living from war. This is the reason why it is correct to speak of war as a social institution, and also the reason why war persists, although everyone realizes that it is the cause of much of the suffering that inflicts humanity. We know that war is madness, but it persists. We know that it threatens the future survival of our species, but it persists, entrenched in the attitudes of historians, newspaper editors and television producers, entrenched in the methods by which politicians finance their campaigns, and entrenched in the financial power of arms manufacturers—entrenched also in the ponderous and costly hardware of war, the fleets of warships, bombers, tanks, nuclear missiles and so on.

Science cannot claim to be guiltless: In Eisenhower's farewell address, he warned of the increasing power of the industrial-military complex—a threat to democratic society. If he were making the same speech today, he might speak of the industrial-military-scientific complex. Since Hiroshima, we have known that new knowledge is not always good. There is a grave danger that nuclear weapons will soon proliferate to such an extent that they will be available to terrorists and even to the mafia. Chemical and biological weapons also constitute a grave threat. The eradication of smallpox in 1979 was a triumph of medical science combined with international cooperation. How sad it is to think that military laboratories cultivate smallpox and that the disease may soon be reintroduced as a biological weapon!

The institution of war seems to be linked to a fault in human nature—to our tendency to exhibit altruism towards members of our own group but aggression towards other groups if we perceive them to be threatening our own community. This tendency, which might be called "tribalism", was perhaps built into human nature by evolution during the long prehistory of our species, when we lived as hunter-gatherers in small genetically homogeneous tribes, competing for territory on the grasslands of Africa. However, in an era of nerve gas and nuclear weapons, the anachronistic behavior pattern of tribal altruism and intertribal aggression now threatens our survival.

Fortunately, our behavior is only partly determined by inherited human nature. It is also, and perhaps to a larger extent, determined by education and environment; and in spite of all the difficulties just mentioned, war has been eliminated locally in several large regions of the world. Taking these regions as models, we can attempt to use the same methods to abolish war globally.

For example, war between the Scandinavian nations would be unthinkable today, although the region once was famous for its violence. Scandinavia is especially interesting as a model for what we would like to achieve globally, because it is a region in which it has been possible not only to eradicate war, but also poverty; and at the same time, death from infectious disease has become a rarity in this region.

If we consider the problem of simultaneously eliminating poverty, war and frequent death from infectious disease, we are led inevitably to the problem of population stabilization. At the time when poverty, disease and war characterized Scandinavia, the average fertility in the region was at least six children per woman-life. Equilibrium was maintained at this high rate of fertility, because some of the children died from disease without leaving progeny, and because others died in war. Today, poverty and war are gone from the Nordic countries, and the rate of premature death from infectious disease is very low. The simultaneous elimination of poverty, disease and war would have been impossible in Scandinavia if the rate of fertility had not fallen to the replacement level. There would then have been no alternative except for the population to grow, which it could not have continued to do over many centuries without environmental degradation, bringing with it the recurrence of poverty, disease and war.

In Scandinavia today, democratic government, a high level of education, economic prosperity, public health, high social status for women, legal, economic and educational equality for women, a low birth rate, and friendly cooperation between the nations of the region are mutually linked in loops of cause and effect. By contrast, we can find other regions of the world where low status of women, high birth rates, rapidly increasing population, urban slums, low educational levels, high unemployment levels, poverty, ethnic conflicts and the resurgence of infectious disease are equally linked, but in a vicious circle. The three age-old causes of human suffering, poverty, infectious disease and war are bound together by complex causal relationships involving also the issues of population stabilization and woman's rights. The example of Scandinavia shows us that it is possible to cure all these diseases of society; but to do so we must address all of the problems simultaneously.

Abolition of the institution of war will require the construction of structures of international government and law to replace our present anarchy at the global level. Today's technology has shrunken the distances which once separated nations; and our present system of absolutely sovereign nation-states has become both obsolete and dangerous.

Professor Elie Kedourie of the University of London has given the following definition of nationalism:

> . . . a doctrine invented in Europe at the beginning of the 19th century. It pretends to supply a criterion for the determination of the unit of population proper to enjoy a government exclusively its own, for the legitimate exercise of power in the state, and for the right organization of a society of states. Briefly, the doctrine holds that humanity is naturally divided into nations, that nations are known by certain characteristics which can be ascertained, and that the only legitimate type of government is national self-government.

A basic problem with this doctrine is that throughout most of the world, successive waves of migration, conquest and intermarriage have left such a complicated ethnic mosaic that attempts to base political divisions on ethnic homogeneity often meet with trouble. In Eastern Europe, for example, German-speaking and Slavic-speaking peoples are mixed together so closely that the Pan-German and Pan-Slavic movements inevitably clashed over the question of who should control the regions where the two populations lived side by side. This clash was one of the main causes of the First World War.

Similarly, when India achieved independence from England, a great problem arose in the regions where Hindus and Muslims lived side by side; and even Gandhi was unable to prevent terrible violence from taking place between the two communities. This problem is still present, and it has been made extremely dangerous by the acquisition of nuclear weapons by India and Pakistan.

More recently, nationalist movements in Asia and Africa have derived their force and popularity from a reaction against the years of European political and economic domination. Thus, at first sight, they seem to deserve our sympathy and support. However, in building states, the new nationalists have often used hate for outsiders as mortar. For example, Israel is held together by hostility towards its Arab neighbors, while the Pan-Arab movement is held together by hostility towards Israel; and in this inflamed political climate of mutual fear and hatred, even clandestine nuclear weapons appear to either side to be justified.

A basic problem rooted in nationalist mythology exists in the concept of sanctions, which treat nations as if they were individuals. We punish nations as a whole by sanctions, even when only the leaders are guilty, even though the burdens of the sanctions often fall most heavily on the weakest and least guilty of the citizens, and even though sanctions often have the effect of uniting the citizens of a country behind the guilty leaders.

It is becoming increasingly clear that the concept of the absolutely sovereign nation-state is an anachronism in a world of thermonuclear weapons, instantaneous communication, and economic interdependence. Probably our best hope for the future lies in developing the United Nations into a World Federation. The strengthened United Nations should have a legislature with the power to make laws which are binding on individuals, and the ability to arrest and try individual political leaders for violations of these laws. The World Federation should also have the military and legal powers necessary to guarantee the human rights of ethnic minorities within nations.

A strengthened UN would need a reliable source of income to make the organization less dependent on wealthy countries, which tend to give support only to those interventions of which they approve. A promising solution to this problem is the so-called "Tobin tax", named after the Nobel-laureate economist James Tobin of Yale University. Tobin proposed that international currency exchanges should be taxed at a rate between 0.1 and 0.25 percent. He believed that even this extremely low rate of taxation would have the effect of damping speculative transactions, thus stabilizing the rates of exchange between currencies. When asked what should be done with the proceeds of the tax, Tobin said, almost as an afterthought, "Let the United Nations have it". The volume of money involved in international currency transactions is so enormous that even the tiny tax proposed by Tobin would provide the World Federation with between 100 billion and 300 billion dollars annually. By strengthening the activities of various UN agencies, such as WHO, UNESCO and FAO, the additional income would add to the prestige of the United Nations and thus make the organization more effective when it is called upon to resolve international political conflicts.

A federation is, by definition, a limited union of states, where the federal government has the power to make laws which are binding on individuals, but where the laws are confined to interstate matters, and where all powers not expressly delegated to the federal government are reserved for the several states. In other words, in a federation, each of the member states runs its own internal affairs according to its own laws and customs; but in certain agreed-on matters, where the interests of the states overlap, authority is specifically delegated to the federal government.

For example, if the nations of the world considered the control of narcotics to be a matter of mutual concern; if they agreed to set up a commission with the power to make laws preventing the growing, refinement and distribution of harmful drugs, and with the power to arrest individuals for violating those laws, then we would have a world federation in the area of narcotics control.

If, in addition, the world community considered terrorism to be a matter of mutual concern; if an international commission were also set up with the power to make global antiterrorist laws, and to arrest individuals violating those laws, then we would have a world federation with somewhat broader powers.

If the community of nations decided to give the federal authority the additional power to make laws defining the rights and obligations of multinational corporations, and the power to arrest individuals violating those laws, then we would have a world federation with still broader powers; but these powers would still be carefully defined and limited.

In 1998, in Rome, representatives of 120 countries signed a statute establishing a Permanent International Court, with jurisdiction over war crimes and genocide. Four years were to pass before the necessary ratifications were gathered, but by Thursday, April 11, 2002, 66 nations had ratified the Rome agreement—six more than the 60 needed to make the court permanent. The jurisdiction of the Permanent International Court is at present limited to a very narrow class of crimes. The global community will have a chance to see how the court works in practice, and in the future, the community may decide to broaden its jurisdiction. In setting up a federation, the member states can decide which powers they wish to delegate to it; and all powers not expressly delegated would be retained by the individual states.

We are faced with the problem of constructing a new world order which will preserve the advantages of local self-government while granting certain carefully-chosen powers to larger regional or global authorities. Which things should be decided locally, or regionally, and which globally?

In the future, overpopulation and famine are likely to become increasingly difficult and painful problems in several parts of the world. Since various cultures take widely different attitudes towards birth control and family size, the problem of population stabilization seems to be one which should be solved locally. At the same time, aid for local family planning programs, as well as famine relief, might appropriately come from global agencies, such as WHO and FAO. With respect to large-scale migration, it would be unfair for a country which has successfully stabilized its own population, and which has eliminated poverty within its own borders, to be forced to accept a flood of migrants from regions of high fertility. Therefore the extent of immigration should be among the issues to be decided locally.

Security, and controls on the manufacture and export of armaments will require an effective authority at the global level. It should also be the responsibility of the international community to intervene to prevent gross violations of human rights. Since the end of the Cold War, the United Nations has more and more frequently been called upon to send armed forces to troubled parts of the world. In many instances, these calls for UN intervention have been prompted by clear and atrocious violations of human rights, for example by "ethnic cleansing" in Bosnia and by genocide in Rwanda. In the examples just named, the response of the United Nations would have been much more effective, and many lives would have been saved, if the action which was finally taken had come sooner. Long and complex diplomatic negotiations were required to muster the necessary political and physical forces needed for intervention, by which time the original problems had become much more severe. For this reason, it has been suggested that the UN Secretary General, the Security Council and the General Assembly ought to have at their disposal a permanent, highly trained and highly mobile emergency force, composed of volunteers from all nations. Such an international police force would be able to act rapidly to prevent gross violations of human rights or other severe breaches of international law.

In evaluating the concept of an international police force directly responsible to the United Nations, it is helpful to examine the way in which police act to enforce laws and to prevent violence and crime at local and national levels.

Within a community which is characterized by good government, police are not highly armed, nor are they very numerous. Law and order are not maintained primarily by the threat of force, but by the opinion of the vast majority of the citizens that the system of laws is both just and necessary. Traffic stops when the signal light is red and moves when it is green whether or not a policeman is present, because everyone understands why such a system is necessary.

Nevertheless, although the vast majority of the citizens in a well-governed community support the system of laws and would never wish to break the law, we all know that the real world is not heaven. The total spectrum of human nature includes evil as well as a good. If there were no police at all, and if the criminal minority were completely unchecked, every citizen would be obliged to be armed. No one's life or property would be safe. Robbery, murder and rape would flourish.

Within a society with a democratic and just government, whose powers are derived from the consent of the governed, a small and lightly armed force of police is able to maintain the system of laws. One reason why this is possible has just been mentioned—the force of public opinion. A second reason is that the law acts on individuals. Since obstruction of justice and the murder of policemen both rank as serious crimes, an individual criminal is usually not able to organize massive resistance against police action.

Edith Wynner, one of the pioneers of the World Federalist movement, lists the following characteristics of police power in a well-governed society:

> 1. A policeman operates within a framework of organized government having legislative, executive and judicial authority operating on individuals. His actions are guided by a clearly stated criminal code that has the legislative sanction of the community. Should he abuse the authority vested in him, he is subject to discipline and court restraint.
>
> 2. A policeman seeing a fight between two men does not attempt to determine which of them is in the right and then help him beat up the one he considers wrong. His function is to restrain violence by both, to bring them before a judge who has authority to determine the rights of the dispute, and to see that the court's decision is carried out.
>
> 3. In carrying out his duties, the policeman must apprehend the suspected individual without jeopardizing either the property or the lives of the community where the suspect is to be arrested. And not only is the community safeguarded against destruction of property and loss of life but the rights of the suspect are also carefully protected by an elaborate network of judicial safeguards.

Edith Wynner also discusses the original union of the thirteen American colonies, which was a confederation, analogous to the present United Nations. This confederation was found to be too weak, and after eleven years it was replaced by a federation, one of whose key powers was the power to make and enforce laws which acted on individuals. George Mason, one of the architects of the federal constitution of the United States, believed that "such a government was necessary as could directly operate on individuals, and would punish those only whose guilt required it", while James Madison (another drafter of the US federal constitution) remarked that the more he reflected on the use of force, the more he doubted "the practicability, the justice and the efficacy of it when applied to people collectively, and not individually". Finally, Alexander Hamilton, in his "Federalist Papers", discussed the confederation with the following words: "To coerce the states is one of the maddest projects that was ever devised . . . Can any reasonable man be well disposed towards a government, which makes war and carnage the only means of supporting itself—a government that can exist only by the sword? Every such war must involve the innocent with the guilty. This single consideration should be enough to dispose every peaceable citizen against such a government . . . What is the cure for this great evil? Nothing, but to enable the . . . laws to operate on individuals, in the same manner as those of states do."

The United Nations is at present a confederation rather than a federation, and thus it acts by attempting to coerce states, a procedure which Alexander Hamilton characterized as "one of the maddest projects that was ever devised". Whether this coercion takes the form of economic sanctions, or whether it takes the form of military intervention, the practicability, the justice and the efficacy of the UN's efforts are hampered because they are applied to people collectively and not individually. It is obvious that the United Nations' actions to stop aggression of one state against another in the Korean War and in the Gulf War fail to match the three criteria for police action listed above. What is the cure for this great evil? "Nothing", Hamilton tells us, "but to enable the laws to act on individuals, in the same manner as those of states do."

Historically, confederations have always proved to be too weak; but federations have on the whole been very successful, mainly because a federation has the power to make laws which act on individuals. At the same time, a federation aims at leaving as many powers as possible in the hands of local authorities. Recent examples of federations include the United States of America, the United States of Brazil, the United States of Mexico, the United States of Venezuela, the Argentine Nation, the Commonwealth of Australia, the Dominion of Canada, the Union of South Africa, Switzerland, the Union of Soviet Socialist Republics and the European Federation. Thus we are rich in historical data on the strengths and weaknesses of federations, and we can make use of this data as we attempt to construct good government at the global level.

Looking towards the future, we can perhaps foresee a time when the United Nations will have been converted to a federation and given the power to make international laws which are binding on individuals. Under such circumstances, true international law enforcement will be possible, incorporating all of the needed safeguards for lives and property of the innocent. One can hope for a future world where the institution of war will be abolished, and where public opinion will support international law to such an extent that a new Hitler or a future Melosovic will not be able to organize large-scale resistance to arrest—a world where international law will be seen by all to be just, impartial and necessary—a well-governed global community within which each person will owe his or her ultimate loyalty to humanity as a whole.

Besides a humane, democratic and just framework of international law and governance, we urgently need a new global ethic,—an ethic where loyalty to family, community and nation will be supplemented by a strong sense of the brotherhood of all humans, regardless of race, religion or nationality. Schiller expressed this feeling in his "Ode to Joy", which Beethoven adopted for the last movement of his Ninth Symphony. Hearing Beethoven's music and Schiller's words, most of us experience an emotion of resonance and unity with its message: All humans are brothers and sisters—not just some—all! It is almost a national anthem of humanity. The feelings which the music and words provoke are similar to patriotism, but broader. It is this sense of a universal human family, which we need to cultivate in education, the mass media, and religion.

Educational reforms are urgently needed, particularly in the teaching of history. As it is taught today, history is a chronicle of power struggles and war, told from a biased national standpoint. Our own race or religion is superior; our own country is always heroic and in the right.

We urgently need to replace this indoctrination in chauvinism by a reformed view of history, where the slow development of human culture is described, giving adequate credit to all those who have contributed. Our modern civilization is built on the achievements of ancient cultures. China, India, Mesopotamia, ancient Egypt, Greece, the Islamic world, Christian Europe, and Jewish intellectual traditions all have contributed. Potatoes, corn and squash are gifts from the American Indians. Human culture, gradually built up over thousands of years by the patient work of millions of hands and minds, should be presented to students of history as a precious heritage—far too precious to be risked in a thermonuclear war.

In the teaching of science too, reforms are needed. Graduates in science and technology should be conscious of their responsibilities. They must resolve never to use their education in the service of war, or in any way which might be harmful to society or to the environment.

In modern societies, mass media play an extremely important role in determining behavior and attitudes. This role can be a negative one when the media show violence and enemy images, but if used constructively, the mass media can offer a powerful means for creating international understanding. If it is indeed true that tribalism is part of human nature, it is extremely important that the mass media be used to the utmost to overcome the barriers between nations and cultures. Through increased communication, the world's peoples can learn to accept each other as members of a single family.

Finally, let us turn to religion, with its enormous influence on human thought and behavior. Christianity, for example, offers a strongly stated ethic, which, if practiced, would make war impossible. In Mathew, the following passage occurs:

"Ye have heard it said: Thou shalt love thy neighbor and hate thy enemy. But I say unto you: Love your enemies, bless them that curse you, do good to them that hate you, and pray for them that spitefully use you and persecute you."

This seemingly impractical advice—that we should love our enemies—is in fact of the greatest practicality, since acts of unilateral kindness and generosity can stop escalatory cycles of revenge and counter-revenge such as those which characterize the present conflict in the Middle East and the recent troubles of Northern Ireland. However, Christian nations, while claiming to adhere to the ethic of love and forgiveness, have adopted a policy of "massive retaliation", involving systems of thermonuclear missiles whose purpose is to destroy as much as possible of the country at which the retaliation is aimed. It is planned that entire populations shall be killed in a "massive retaliation", innocent children along with the guilty politicians. The startling contradiction between what the Christian nations profess and what they do was obvious even before the advent of nuclear weapons, at the time when Leo Tolstoy, during his last years, was exchanging letters with a young Indian lawyer in South Africa. In one of his letters to Gandhi, Tolstoy wrote:

> . . . The whole life of the Christian peoples is a continuous contradiction between that which they profess and the principles on which they order their lives—a contradiction between love accepted as the law of life, and violence, which is recognized and praised, acknowledged even as a necessity . . .
>
> This year, in the spring, at a Scripture examination at a girls' high school in Moscow, the teacher and the bishop present asked the girls questions on the Commandments, and especially on the sixth. After a correct answer, the bishop generally put another question, whether murder was always in all cases forbidden by God's law; and the unhappy young ladies were forced by previous instruction to answer "Not always"—that murder was permitted in war and in the execution of criminals. Still, when one of these unfortunate young ladies (what I am telling is not an invention but a fact told to me by an eye witness) after her first answer, was asked the usual question, if killing was always sinful, she, agitated and blushing, decisively answered 'Always', and to the usual sophisms of the bishop, she answered with decided conviction that killing was always forbidden in the Old Testament and forbidden by Christ, not only killing but every wrong against a brother. Notwithstanding all his grandeur and arts of speech, the bishop became silent and the girl remained victorious.

As everyone knows, Gandhi successfully applied the principle of non-violence to the civil rights struggle in South Africa, and later to the political movement, which gave India its freedom and independence. The principle of non-violence was also successfully applied by Martin Luther King, and by Nelson Mandela. It is perhaps worthwhile to consider Gandhi's comment on the question of whether the end justifies the means: "The means may be likened to a seed," Gandhi wrote, "and the end to a tree; and there is the same inviolable connection between the means and the end as there is between the seed and the tree." In other words, a dirty method produces a dirty result; killing produces more killing; hate leads to more hate. Everyone who reads the newspapers knows that this is true. But there are positive feedback loops as well as negative ones. A kind act produces a kind response; a generous gesture is returned; hospitality results in reflected hospitality. Buddhists call this principle of reciprocity "the law of karma".

The religious leaders of the world have the opportunity to contribute importantly to the solution of the problem of war. They have the opportunity to powerfully support the concept of universal human brotherhood, to build bridges between religious groups, to make intermarriage across ethnic boundaries easier, and to soften the distinctions between communities. If they fail to do this, they will have failed humankind at a time of crisis.

It is useful to consider the analogy between the institution of war and the institution of slavery. We might be tempted to say, "There has always been war, throughout human history; and war will always continue to exist." As an antidote for this kind of pessimism, we can think of slavery, which, like war, has existed throughout most of recorded history. The cultures of ancient Egypt, Greece and Rome were all based on slavery, and, in more recent times, 13 million Africans were captured and forced into a life of slavery in the New World. Slavery was as much an accepted and established institution as war is today. Many people made large profits from slavery, just as arms manufacturers today make enormous profits. Nevertheless, in spite of the weight of vested interests, slavery has now been abolished throughout most of the world.

Today we look with horror at drawings of slave ships, where human beings were packed together like cordwood; and we are amazed that such cruelty could have been possible. Can we not hope for a time when our descendants, reading descriptions of the wars of the twentieth century, will be equally amazed that such cruelty could have been possible? If we use them constructively, the vast resources now wasted on war can initiate a new era of happiness and prosperity for the family of man. It is within our power to let this happen. The example of the men and women who worked to rid the world of slavery can give us courage as we strive for a time when war will exist only as a dark memory fading into the past.

Bibliography

Axelrod, R., *The Evolution of Cooperation* (New York: Basic Books 1984).

Babbst, D.V., et al., *Accidental Nuclear War: The Growing Peril* (Dundas, Ontario: Peace Research Institute, 1984).

Chivian, E., et al. (eds.) (International Physicians for the Prevention of Nuclear War) *Last Aid: The Medical Dimensions of Nuclear War* (San Francisco: W.H. Freeman, 1982).
Freud, S., *Warum Krieg? Das Bild vom Feind* , Arbeitsgem. Friedenspedagogik (1983).
Guenther, N.A., *Children and the Threat of Nuclear War: The Growing Peril* (Dundas, Ontario: Peace Research Institute, 1984).
Hinde, R.A., *The Biological Basis of Human Social Behavior* (New York: McGraw Hill, 1977).
Hinde, R.A., *Towards Understanding Human Relationships* (London: Academic Press, 1979).
Kahert, M., et al., (eds.) *Children and War* (Finland: Peace Union of Finland,1983).
Kedourie, E., *Nationalism in Asia and Africa* (London: Frank Cass & Co., 1970).
Kamenka, E., (ed.) *Nationalism* (London: Edward Arnold, 1976).
Koestler,, A, "The Urge to Self-Distruct" in *The Place of Values in a World of Facts*, A.Tiselius and S. Nielsson, (eds.), (New York: Wiley, 1970).
Levine, R.A., Campbell, D.T., *Ethnocentrism: Theories of Conflict, Ethnic Attitudes and Group Behavior* (New York: Wiley, 1972).
Mahler, H., *World Health is Indivisible* (Geneva: World Health Organization, 1978).
Schear, J., (ed.), *Nuclear Weapons Proliferation and Nuclear Risk* (London: Gower, 1984).
Wright, Q., *A Study of War* (Chicago: Chicago University Press, 1965).
Wynner, E., *Federal Government in Maximum Terms: Proposals for United Nations Charter Revision* (Afton, NY: Fedonat Press, 1954).
Zahn-Waxler, C., *Altruism and Aggression: Biological and Social Origins* (Cambridge: Cambridge University Press, 1986).

Sean Howard

Sean Howard *is adjunct professor of political science at Cape Breton University, Nova Scotia, Canada. He received his PhD from the Department of Peace Studies, University of Bradford, UK, and from 1992 to 2003 worked as a researcher on nuclear disarmament and non-proliferation issues for a number of non-governmental organizations. After moving to Canada in 1999, his research interests expanded to include the military and environmental threats of new technologies, particularly nanotechnology, and the growing relevance of indigenous perspectives in the contemporary search for peace and justice. His work on a range of topics has appeared in journals and magazines including* Resurgence, Peace Magazine *and* Disarmament Diplomacy.

VERY DIFFERENT BUTTERFLIES:

The Scope for Deep Complementarity Between Western and Native American Science[1]

Sean Howard

. . . the same chrysalis, according as it was submitted to cold or heat, gave rise to very different butterflies, which had long been regarded as independent species, *Vanessa levana* and *Vanessa prorsa* . . .

Henri Bergson, *Creative Evolution*, 1907[2]

The world is already enchanted—*really, here, now*! Every tree, every rock, every star, and space itself has awareness and living energy. We have the ability to see this. And there *are* patterns of energy that can be clearly felt but usually not seen. Call them gods, diamons, fairies, angels, *dralas*, even meaningful coincidence; call them what you will. And *that* story of unceasing awareness could have been told to us as we grew up *along with* the tiny, narrow story of the Dead World. And it too could have been told in the name of science.

Jeremy Hayward, *Letters to Vanessa*, 1997[3]

I. Introduction

Complementarity and Western Science: The Incomplete Revolution

In 1927, the Danish physicist and Nobel Laureate Niels Bohr proposed the simple yet profound principle of 'complementarity' to guide description of physical reality at the atomic level. Complementarity was a conceptual and philosophical response to the technical and mathematical proof—afforded a few months earlier by Werner Heisenberg's 'uncertainty principle'—that the velocity and position of a subatomic particle can never be simultaneously determined. Necessarily relinquishing the search for a single coherent picture—a self-contained, exhaustive description—of the atom, Bohr suggested the virtue of an *inclusive* overview, encompassing the mutually *exclusive* descriptions of atomic activity (e.g. electron as particle, electron as wave) derived from experiments. Hitherto, the objective of atomic physicists had been to *integrate* conflicting signals from the atomic realm into a definitive interpretation. Henceforth, the objective was *integrative*: to deepen what Bohr called the "harmonious comprehension" of that realm through critical appreciation of its necessarily distinctive manifestations.[4]

This shift, radical though it is, does not imply a rejection of natural reality, but rather a recognition of its *uncapturable* essence, its irreducibility to anything other than a partial representation; valuable, complementary to other partial descriptions, but necessarily incomplete. As Bohr insisted: "It is wrong to think that the task of physics is to find out about nature. Physics concerns what we can say about nature."[5] From this perspective, the 'curse' of contradictory descriptions becomes a blessing—precisely because such descriptions *are* contra-dictions, different ways of speaking about nature.

Crucial in shaping this new approach was Bohr's recognition of the decisive role and radical implications of *wholeness* at the atomic level: not only the indivisible pulses (quanta) of energy exchanged within and between atoms, but also the *whole situation* created to investigate atomic 'objects' of inquiry in the laboratory. As the uncertainty principle demonstrated, at the atomic level the activity inseparable from any act of observation necessarily changes the nature of the activity observed. A microscope can only observe electrons, for example, *interactively*: by emitting particles of light (photons) which radically alter—*become part of*—the energy and trajectory of the particles they strike. It was this basic "observational problem", in Bohr's words—the potentially impassable "circumstance that the interaction between atomic objects and measuring instruments forms an integral part of quantum phenomena"—that necessitated "the complementary mode of description", with its explicit acknowledgement of the "necessity of taking into account the conditions under which experience is obtained."[6]

Bohr's immediate goal in developing complementarity was, thus, twofold: to save physics from a false objectivity—discovering "how nature is"—while opening the path to a meaningful interactivity—discovering what "we can say about" those aspects of the natural world we engage.

Together with Heisenberg and other close colleagues, however, Bohr gradually came to see complementarity as a principle with a far wider field of application than atomic physics. With regard to biology, for example, Bohr saw a parallel between "the essential non-analyzability of atomic stability in mechanical terms"—that is, the irreducible *unity* of atomic phenomena—and the "existence of life itself". "Life," he concluded, "should be considered, both as regards its definition and observation, as a basic postulate of biology, not susceptible of further analysis, in the same way as the existence of the quantum of action, together with the ultimate atomicity of matter, forms the elementary basis of atomic physics."[7] The task of biology, in other words, is not to define what life *is*, but to explore the different patterns streaming from that untraceable Source.

The presumed "ultimate" nature of this "atomicity"—the decisive but indefinable role of wholeness in natural phenomena— extends for Bohr into the heart of 'human nature' itself, forming the "elementary basis" of all our "modes of expression", including the scientific.[8] Bohr "felt," as the physicist F. David Peat puts it, that complementarity "was basic to human consciousness and to the way the mind works":

> Until the twentieth century, science had dealt in the certainties of Aristotelian logic: 'A thing is either A or not-A.' Now it was entering a world in which something can be 'both A and not-A.' Rather than . . . drawing a single map that corresponds in all its features to the external world, science was having to produce a series of maps showing different features, maps that never quite overlap.[9]

But, crucially for Bohr, science is also *part* of the territory it seeks to chart. While it may now produce a plurality of representations, that plurality itself can be seen as *one* dimension of the illimitable, ever-unfolding 'meta map' of the human journey. Seen in this inclusive, integrative light, science enjoys not a superior or hierarchical but a complementary relationship with art, music, poetry, philosophy, religion, mysticism, etc. A poem is insusceptible to scientific analysis; mathematical formalism is irreducible to poetic metaphor. But though all these partial ways of speaking about nature are contra-dictions—employing incommensurate terms of engagement with reality—each stems from the same unity, the indivisible *root* of living diversity itself. What other way, Bohr asks, *except* diversely, can natural diversity be represented and explored?

In the wake of World War II and the dawn of the atomic age, Bohr and his supporters saw in complementarity the vital clue—unity-in-diversity—to the new human order necessitated by the destructive applications of modern science. Although, in Heisenberg's words, "scientific concepts cover always only a very limited part of reality," it is precisely this recognition, or newfound humility, that is of value in the search for "a calmer kind of evolution", a very different history: "through its openness to all kinds of concepts . . . [modern physics] raises the hope that . . . many cultural traditions may live together and may combine different human endeavors into a new kind of balance between thought and deed, between activity and meditation."[10]

For many Western scientists, however, complementarity remains a strange and all-too-subtle—almost an *un*scientific—concept. The British physicist Paul Dirac, for example, commented in 1963 that he "didn't altogether like it" because "it doesn't provide you with any equations which you didn't have before."[11] According to the historian of science Helge Krage, the bulk of practicing physicists—particularly in North America, with its traditionally pragmatic emphasis on the *results* of scientific inquiry—agreed with Dirac and simply "ignored Bohr's philosophy."[12] The assessment is echoed by the American writer Robert Pirsig: "most physicists use the mathematics of the quantum theory with complete confidence and completely ignore the philosophy."[13]

From the opposite direction, Albert Einstein led the *philosophical* charge against complementarity, insisting that science remain committed to producing a single, self-sufficient map of physical reality—the elusive 'grand unified field theory', or 'theory of everything', which would, in the words of Stephen Hawking, "be the ultimate triumph of human reason—for then we would know the mind of God."[14]

Ironically enough, almost all the scientists pursuing this Holy Grail, while unquestioningly accepting the technical and interpretative parameters established by complementarity and the uncertainty principle, either ridicule or ignore Einstein's rejection of quantum theory. The result, as Peat and physicist David Bohm note, is that they simultaneously "decry" and "actually think like" Einstein while both praising Bohr's science and "tacitly ignoring" his "teaching."[15]

This unresolved tension—a distinctly *uncomplementary* contra-diction within modern physics—is related, I believe, to the broader failure of the *Western* scientific worldview to acknowledge the legitimacy, or even the existence, of *other* scientific worldviews. Perhaps surprisingly, this failure was common across the spectrum of debate over complementarity: for Bohr and Heisenberg as much as Einstein and Dirac, science *meant* the language and methods of rational, enlightened inquiry into physical reality bequeathed to the world by the European Age of Reason. While complementarity, from the perspective of its 'founding fathers', implied the need for a significant revision of that tradition, requiring that science set itself on an equal footing with other modes of inquiry, it did not imply or require the expansion of scientific self-awareness from the Western to the global, i.e. the inclusion of non-Enlightenment traditions. The 'other modes' of inquiry in question were, by definition, *non*-scientific: science was, precisely, a mode of inquiry incapable of entering into a complementary relationship with itself. While modern physics *had* now to accommodate complementary descriptions of natural phenomena, it could *never* entertain complementary approaches *to* physics.

This view is seen most clearly in Bohr's assumptions about the central role of "language"—how we *have* to speak of nature—in the scientific representation of reality. "During the development of quantum physics," he wrote, "entailing so many surprises, we have time and again been reminded of the difficulties of orienting ourselves in a domain of experience far from that to the description of which our means of expression are adapted."[16] But the "problems of language here are really serious," as Heisenberg commented, because not only are "the concepts of ordinary language" inadequate ("inaccurate and only vaguely defined"[17]) to the task of a scientific description of the atom: they are also *indispensable* to any such account. Not only is exploration of the quantum microworld only possible using macroscale equipment based on the principles and processes of classical physics; interpretation of experimental evidence from the quantum realm is only possible using classical concepts of time, space, matter, energy, etc. The extraordinary and unfamiliar (quantum) is only rendered explicable in terms of the ordinary and everyday (classical); the translation is as incomplete as it is unavoidable.

Thus, the most contentious issues raised by the quantum theory, Heisenberg wrote, "concern the language rather than the facts, since it belongs to the concept 'fact' that it can be described in ordinary language."[18] These 'problems' are 'solved' by complementarity not through provision of a new language but acceptance of the provisional nature of all explanation: by seeking to increase what we can say about nature in terms ultimately inapplicable to its extraordinary, unanalyzable and—from the viewpoint of classical theory—entirely unanticipated wholeness of action.

But is it true that *all* language is restricted in the same way; that there is one basic set of concepts, and thus one broad linguistic organization of conceptual thought? If so, then the 'fact' that 'science' arose in a certain place at a certain time cannot be fundamentally related to the internal nature of a particular, culturally-specific *kind* of speech: the same way of *thinking* would be natural, allowing for differences of *degree*, wherever people enjoyed a developed investigative vocabulary. *If*, however, different *modes* of scientific inquiry, based on distinct conceptual and linguistic terms of engagement with the natural world, were possible, then the advocates of complementarity, in fidelity to their own inclusive principles, would be obliged to reject the hypothesis of an integrated scientific *approach* as decisively as they aborted the search, within *their* approach, for a single, final-word description of reality.

What, for example, if a language of scientific inquiry existed in which the wholeness of action exhibited at the quantum level was entirely *anticipated*, regarded not as shattering but confirmatory of an existing set of assumptions and beliefs, around which key concepts and methods had long been structured?

In the decades since the 'quantum revolution', there have been numerous attempts to argue that such complementary worldviews do exist, offering insights and experiences far too valuable and relevant for Western science to either ignore or relegate to a sub-scientific (and exploitable) realm of 'knowledge'.

Perhaps the most influential of these attempts emerged from modern physics itself: Fritjof Capra's *The Tao of Physics* (1976). As the book's subtitle—*An Exploration of the Parallels between Modern Physics and Eastern Mysticism*—suggests, Capra was at pains to stress the resonance *between*, rather than the reconcilability *of*, Western and Eastern approaches to nature and knowledge. New connections can only be created, new relationships established, in the space *between* different lines of inquiry—in the same way that the fundamental creative Unity of the Tao, the Universal Way, is a balance maintained by what Capra calls the "dynamic interplay" of opposites.[19] As symbolized perfectly by the yin-and-yang circle of Chinese philosophy, the wheel of being—whether in the cosmic macrocosm or the individual microcosm—is kept in motion by the creative tension within it: and this motion is itself the basic energy or spirit in-forming (and irreducible to) its partial manifestations. To assume the constant, decisive influence of such interactivity at all levels of natural reality—to *look for* these oppositional indicators (and generators) of a unified field—is to take as your *starting point* the line marking the *break* between modern and classical Western physics. As Capra writes:

> For a better understanding of this relationship between pairs of classical concepts, Niels Bohr has introduced the notion of complementarity . . . [Bohr] has often suggested that it might be a useful concept also outside the field of physics; in fact, the notion of complementarity proved to be extremely useful 2,500 years ago. It played an essential role in ancient Chinese thought which was based on the insight that opposite concepts stand in a polar—or complementary—relationship to each other.[20]

Capra notes that Bohr "was well aware of the parallel between his concept of complementarity and Chinese thought" and was "deeply impressed by the ancient Chinese notion of polar opposites." In 1947, for example, awarded a knighthood by his native Denmark, Bohr chose as the design of his coat-of-arms "the Chinese symbol of *t'ai-chi* representing the complementary relationship of the archetypal opposites yin and yang", which, "together with the inscription *Contraria sunt complementa* ('Opposites are complementary')," reflected his grateful acknowledgement of "the profound harmony between ancient eastern wisdom and modern Western science."[21]

This very formulation, however, encodes an opposition requiring closer analysis: ancient wisdom *versus* modern science. Why not reformulate the polarity as Chinese versus Western *science*? All scientific approaches to physical reality are rooted in essentially philosophical assumptions, beliefs and convictions. The 'wisdom' shaping physical inquiry, its meta-physical form, is necessarily broader than its content, the body of knowledge *thus derived*. The frame, as Einstein once remarked to Heisenberg, shortly before complementarity reshaped modern physics, is always bigger than the picture: "It is the theory which decides what we can observe."[22]

The objective of complementarity is to expose and thus prevent false choices between aspects of reality—and dimensions of inquiry—better considered as parts of a larger whole. Knowledge and wisdom within a scientific tradition, however, are not complementary but *integral* to each other. The complementary relationship exists, instead, between *different lines and languages* of inquiry, distinct knowledge-wisdom complexes reflecting different historical, social and cultural experience.

It is, thus, legitimate to draw parallels *between* the Chinese and Western 'complexes'—while remembering that the parallelism is possible only because the lines do *not* meet. To polarize these traditions into false opposites—the modern (scientific rationality) versus the ancient (traditional knowledge)—is dangerous, however, because, *from a modern Western perspective*, the comparison ultimately denigrates the Other it creates: 'ancient' equals pre-scientific, non-scientific, even anti-scientific. Instead of a parallel—legitimate distance, rooted in difference—we have a *prejudice*, an illegitimate distortion obscuring underlying similarities. The 'Western' and 'Eastern' traditions, in Capra's broad-brush terms, are *both* sciences precisely because they are *different* sciences. Merely because the *Western* tradition is currently undergoing dramatic change, however, does not entitle it to either subsume or appropriate other approaches. What is needed is not a *merging* but a *meeting* of minds: a recognition of the difference always implied by similarity, the common ground which can only be established *between* camps.

The same principles apply, and the same dangers arise, with regard to the parallels between the Western and Native American traditions. Are we dealing with a complementarity between different sciences, or the complementarity of modern science and traditional knowledge? Can Native American understandings of the natural world—its frame of physical reference—be regarded as a distinct line of scientific development, the expression of a different language of scientific inquiry? Or is 'Native knowledge' simply another legitimate object of scientific inquiry, a pre-scientific language ripe not just for translation but modernization?

The remainder of this paper defends the position that complementary *scientific* descriptions can indeed be derived—mutually-instructive parallels drawn—by examining natural reality through the different lenses afforded by the two (Native and Western) worldviews. By 'scientific', I mean that each 'lens' brings a clear, organized, self-consistent theory into focus via a set of recognized and dependable methods of observation, replication, verification and interpretation. Because of the profound conceptual differences shaping those physical investigations, the modes of inquiry and means of expression deployed cannot be combined into a single, *synthetic* (i.e. artificial, unnatural) approach. Taken in combination, however—as a *syncretic* totality, a resonance sustained by different frequencies—the comprehensiveness of perspective on natural phenomena is enriched.

If the merits of such a harmonic approach were to find acceptance in Western science, the path would stand open to an expansion of complementarity, a deepening of the concept to include not just a plurality of modes of human engagement with nature, of which 'science' is *one*, but a plurality of *scientific* modes of inquiry.

After a general introduction to both the frame-of-reference of Native American science and the picture of reality therein presented, an example of such 'deep complementarity' will be considered—"chaos theory", or "creative energy", the dynamic interplay of flux, field and form in natural systems.

II. Going Deeper

Complementarity and Native American Science

In his 1994 book *Lighting the Seventh Fire*—reissued as *Blackfoot Physics*—David Peat argued that Native American understandings of the natural world should indeed be considered scientific. All the topics to be covered in his study, he writes—"metaphysics and philosophy; the nature of space and time; the connection between language, thought, and perception; mathematics and its relationship to time; the ultimate nature of reality; causality and interconnection; astronomy and the movement of time; healing; the inner nature of animals, rocks, and plants; powers of animation; the importance of maintaining a balanced exchange of energy; of agriculture; of genetics; of considerations of ecology; the connection of the human being to the cosmos; and of the nature of processes of knowing"—can best be "gathered together under the general rubric of Indigenous science, a term I have used following the lead of Pam Colorado, Leroy Little Bear, and others."

Having staked out his position, Peat immediately acknowledges its lonely quality: "While I am comfortable with this term," he adds, "I am also a part of the Western science tribe and I can already sense the kind of objections that its members would make: 'Why do you use the term *science*? Native Americans don't have any science in the real sense of the word. They don't have an ordered system of investigation or rational theories of the universe as we do. Science is a specific and disciplined approach that was developed in the West. Indigenous people have traditions, folklore and mythology.'"[23] This general view is expressed with particular vigor by the biologist Edward O. Wilson:

> I mean no disrespect when I say that pre-scientific people, regardless of their innate genius, could never guess the nature of physical reality beyond the tiny sphere attainable by unaided common sense. Nothing else ever worked, no exercise from myth, revelation, art, trance, or any other conceivable means; and notwithstanding the emotional satisfaction it gives, mysticism, the strongest pre-scientific probe into the unknown, has yielded zero. No shaman's spell or fast upon a sacred mountain can summon the electromagnetic spectrum. Prophets of the great religions were kept unaware of its existence, not because of a secretive god but because they lacked the hard-won knowledge of physics.[24]

Throughout his survey, Peat answers this objection on two levels: by demonstrating the central importance in the Native American tradition of the main attributes, as conventionally defined in the *West*, of a scientific approach—observation, experimentation, prediction, verification, replication, modeling, etc; and by showing that the radical differences in the *conclusions* drawn in the two traditions derive not from the application of superior or inferior techniques of inquiry but rather from the radically different philosophical *premises* involved. What, for example, really *is* the electromagnetic spectrum 'summoned' by Western physics: a complete and natural explanation of the *whole* phenomenon, or *a way of describing aspects of it* natural to, and selected by, a particular approach?

Interestingly, this latter view—the acceptance of the inherent partiality of scientific 'truth'—would not only describe Bohr's position but a basic, guiding tenet of the Native American worldview: the irreducibility of the Great Spirit to any of Its manifestations, of Flux (dynamic wholeness) to the forms It engenders. Set against this modesty, Wilson's supremacism strikes a doubly-harsh, and ironic, tone: in its assertion of the reducibility of natural reality to scientific law, harkening back to a classical, pre-complementarity Western view; in its championing of modern physics, blind to the major philosophical insight of that essentially *disillusioning* breakthrough. Yet this irony is further compounded by Bohr's own anachronistic identification of Western science—a language of investigation—with that very classical phantom 'science itself', the only language of inquiry on offer—from the West, to the world.

An extended explanation for Bohr's failure to cross this threshold is offered in the conclusion. Peat contrasts the *exclusionary reflex*, the either/or 'nature' of Western cultural perception, with the *inclusive capacity* characteristic of the Native American tradition:

> Our Western concept of nature is based on an evolutionary model. Left to the natural forces around them, things will 'progress', getting better and better. Going along with this worldview is the need, when faced with alternatives, to decide which one is 'better' than the others. It goes without saying that when it comes to other people's cultures we are generally the ones doing the measuring, and are supplying the yardstick as well! If two systems exist, both of which claim to be sciences, our natural tendency is to compare them, like the latest model autos, and see which one comes out on top.[25]

Thus, despite complementarity and the uncertainty principle, confronted with two radically different interpretations of the same phenomenon the predominant Western tendency is still to *choose* between them: judgment—'true' or 'false', scientific or *not*—is passed in the case of sacred energy *versus* electromagnetism, natural selection *or* creative evolution, etc. Peat recalls being asked by Native American scientist Sa'ke'j Henderson, during a discussion of the qualities of medicinal plants, "what I thought a molecule was":

> I offered him an explanation from modern science, that a molecule is a geometrical arrangement of atoms. Of course, he knew this kind of answer, but replied that a molecule was an alliance of spirits, and that when taken up into the body this alliance dissolves and takes up new configurations.[26]

To which Wilson would answer, *wrong, you cannot possibly have obtained such an insight other than by the methods of my tradition, which has produced no evidence of any such 'alliances'.* Note, however, that not only was Henderson familiar with Peat's explanation, he made no effort to refute its *relative* validity, i.e. its *internal* self-consistency. Peat reacted initially by attempting to "bridge the gap between our two ways of thinking by calling on concepts from modern quantum physics": "A molecule is an arrangement of atoms but it can also be represented by a wave function, which is, in a way, a sort of vibration of matter and energy. Maybe it is possible to think of the molecules that make up medicines as patterns of vibrations or more subtle forms of matter energy."[27] It is indeed *possible*, he concluded, but to *whose* way of thinking would such a 'bridging' do justice?

> The more I thought about it the more I realized that 'spirit' cannot really be reduced to our words energy or matter as they are currently understood in Western science. The idea of using plants and herbs to cure sickness may at first sight appear close to our scientific idea of 'medicines', suggesting that the powers of animal and vegetable substances lie in their biologically active molecules. But this cannot be the whole of the story. For example, the way these plants are collected is important, and before the medicine bag is used, the bag must be smudged with the purifying smoke of sage, sweet grass, tobacco, or cedar. Medicine itself has a life of its own. It is the tangible manifestation of alignments with the world of powers, energies, and spirits. Trying to understand how and why a particular plant is 'medicine' stretches our Western paradigm to the limit.[28]

Henderson's approach, Peat suggests, is indicative of a general disposition within the Native American tradition to explore rather than judge, juxtapose rather than integrate, alternative interpretations:

> This [Western] desire to compare, to measure, and to categorize in terms of doing better and worse does not seem to be the natural way of doing things within the Indigenous world. . . . [I]n Labrador, there was a traditional Indigenous way to hunt beaver and one that the European trappers used. Today both methods are used by Native people, and there is no sense that one method is 'better' than the other, or that one should replace the other. Rather, both methods are used, side by side.[29]

Such an approach is, however, philosophic as well as pragmatic. Just as there are different ways of hunting animals, so there are different methods for obtaining reliable and useful knowledge of the natural world. But the *source* of being, the *basis* of both the knowledge derived and the methods employed, remains irreducible to formulation, opaque to the forms of perception it creates. You can capture the beaver, skin and dissect it, but you can never *explain* what you describe, reduce the whole to the parts (material and analytic) at your disposal. And, in fact, even this descriptive capacity is severely circumscribed. To examine any entity (animal, atom) in isolation is to dis-locate it from the flux that gave it form, the dynamic integrity it reflects: animal as manifestation of forest, forest of seasons, seasons of Earth, Earth of Cosmos, Cosmos of Creator, etc. To dissect is to reduce this integral reality to a separable display, divisible in space as 'parts' and time as cause-and-effect. The embedded, vital entity becomes a dead *thing*, a machine; *an object of consuming interest*. But whereas the actual consumption of meat for food in Native traditions is invariably located within a broader, non-reductionist context—an exchange or circling of energies within an interdependent Whole—the danger of unqualified reductionism is that it mistakes its own limits, the determining context of its *model* of inquiry, for those of the world it thus 'comes out on top' of every time.

For Bohr, such a view is blindness, a delusion of interpretative grandeur shattered by the profound evidence of dynamic wholeness at the atomic base of existence. The irreducible 'mysteriousness' of life, in short, is not some superstitious, pre-scientific obstacle in the way of Enlightenment, but a scientific premise from which rational, qualified conclusions can be drawn:

> [W]e should doubtless kill an animal if we tried to carry the investigation of its organs so far that we could tell the part played by the single atoms in vital functions. In every experiment on living organisms there must remain some uncertainty as regards the physical conditions to which they are subjected, and the idea suggests itself that the minimal freedom we must allow the organism will be just large enough to permit it, so to say, to hide its ultimate secrets from us. On this view, the very existence of life must in biology be regarded as an elementary fact that cannot be derived from ordinary mechanical physics. Indeed, the essential non-analyzability of atomic physics in mechanical terms presents a close analogy to the impossibility of a physical or chemical explanation of the peculiar functions characteristic of life.[30]

A science that sets out to 'get to the bottom' of life, or the lifeless mechanisms somehow 'causing' the 'effect' of existence, will necessarily differ, *qualitatively*, from a science seeking to explore the holistic, relational diversity of being and becoming. Western science is, of course, capable of developing in this direction, of reconfiguring itself to take full, consistent account of the holistic implications of complementarity. To achieve and sustain this conceptual resonance with other traditions, however, it will need to overcome not only deep-seated cultural prejudices and assumptions of superiority, but *its own* isolation, its compartmentalized dis-location in modern society. For Bohr, 'science' was naturally complementary to other dimensions of inquiry and experience—art, religion, politics, justice, economics, etc. From the Native American perspective, however, these categorical distinctions are unnatural and insupportable. Peat makes the point well:

> I believe that Indigenous science . . . is a disciplined approach to understanding and knowing, or rather, to the processes of coming to understanding and knowing. It has supporting metaphysics about the nature of reality, deals in systems of relationship, is concerned with the energies and processes within the universe, and provides a coherent scheme and basis for action. On the other hand, it is not *possible* to separate Indigenous science from other areas of life such as ethics, spirituality, metaphysics, social order, ceremony, and a variety of other aspects of daily existence. Thus it can never be a 'branch' or 'department' of knowledge, but rather remains separable from the cohesive whole, from a way of being and of coming-to-being.[31]

Perhaps it was this integral quality of Indigenous science, rather than any methodological or epistemological considerations, that blinded Bohr (and blinds Wilson) to its legitimacy; to, precisely, its *separate* existence, with its own institutions, bureaucracies, practices, hierarchies, terminology, specializations, etc. Taken together, as Peat stresses, these 'indicators' of 'progress' in science form the yardstick by which the 'authenticity' of other traditions are judged—as a rule, the stick with which they are beaten. The significance of Peat's contribution, as a modern physicist seeking to change Western perceptions of science, is thus not primarily in demonstrating that the two approaches have important elements in common—important though it is to challenge the monopolization of key concepts such as 'empiricism', 'observation,' 'experimentation,' and 'verification'—but in recognizing the existence of complementary *gauges* of success. "Until now," he writes toward the end of *Blackfoot Physics*, "I have been asking if Indigenous science is truly a science, and in doing so I assumed that only one yardstick exists that provides a standard set of definitions—Western science." Answering 'yes' to the question thus phrased, however, is "not to question whether Indigenous science exists in its own right":

> As far as Indigenous people are concerned, there is no need for them to justify their spirituality, their traditions, or their science by reference to anything external to their society. Indigenous science does not need to explain itself to anyone. It has no need to compare or authenticate itself against the standard of Western science.[32]

In a concise introductory summation of what he calls the 'Native American paradigm', Leroy Little Bear comments:

> Science has been and can be defined many different ways depending on who is doing the defining. But the one thing that is certain is that 'science' is culturally relative. . . . Scientific facts . . . are as much a product of the observer's human nature as they are of an underlying reality. . . . If science is a search for reality and if science is a search for knowledge at the leading edge of the humanly knowable, then there are 'sciences' other than the Western . . . One of these other sciences is Native American science.[33]

The need for validation is one thing, however, and the desire for constructive dialogue another. "The Native American paradigm," Little Bear writes, "is comprised of and includes ideas of constant motion and flux, existence consisting of energy waves, interrelationships, all things being animate, space/place, renewal, and all things being imbued with spirit." This basic framework is described by the same author as "incomprehensible to most Westerners", principally because of its non-reductionist framework in which "measurement is only one of many factors to be considered."[34]

Most, perhaps, but not all; and not inevitably. Complementarity is based on a recognition of the radical limits of measurement, an admission of the partial validity of quantitative analysis of natural processes; and in this contraction of the concept, new models and ideas, more resonant with Native American understandings, are opening up. In quantum physics, for example, measurement no longer describes the deterministic, cause-and-effect relationships and processes central to classical theory; rather, quantitative indicators are obtained of the *non-deterministic*, probabilistic quality of atomic phenomena. On the macroscale—as we'll note in our consideration of chaos theory—similarly non-deterministic qualities of flux and form are the focus of intense study, with the role of measurement again changing from validation of a predictive reductionist hypothesis to the modeling of the role of unpredictability, spontaneity and randomness in systematic change and evolution. Such measurement, in fact, exists on the border of quantity and quality, posing enormous technical and conceptual challenges. As the mathematician Ian Stewart writes:

> One of the most striking features of recent mathematics has been its emphasis on general principles and abstract structures—on the qualitative rather than the quantitative. The great physicist Ernest Rutherford once remarked that "qualitative is just poor quantitative", but that attitude no longer makes sense. To turn Rutherford's dictum on its head, quantitative is just poor qualitative. . . . We will never understand the growth of a tree or the dunes in the desert if we try to reduce all of nature's freedom to restrictive mathematical schemes. The time is ripe for the development of a new kind of mathematics, one that possesses the kind of intellectual rigor that was the real point of Rutherford's criticism of sloppy qualitative reasoning, but has far more conceptual flexibility. We need an effective mathematical theory of form, which is why I call my dream [of a new math] 'morphomatics'.[35]

In classical Western science, measurement was important not in itself but because of its assumed reflection of the immutable, eternal *laws* underlying, and ultimately determining, natural phenomena. As Leroy Little Bear observes, this assumption is alien to the Native American paradigm: "For Einstein and Western science, creation and existence were made in a certain way by God and will always remain the same; everything and anything in creation just needs to be discovered by humans. Nothing is certain unless it can be referred to as a regular pattern after long-term observation. But, for the Native American, even regularities are subject to change. Native Americans never claim regularities as laws, or as finalities. The only constant is change."[36] This, of course, was precisely why Einstein was philosophically repulsed by the 'new physics'; the acausality, randomness and unpredictability it perceived in the quantum realm seemed to him lawless, anarchic, sacrilegious. But as the iron grip of physical law was relaxed, new, more organic conceptions of the relationship between law and reality became possible. A striking recent example is *The Life of the Cosmos* by the physicist Lee Smolin:

> The laws of nature themselves, like the biological species, may not be eternal categories, but rather the creations of natural processes occurring in time. There will be reasons why the laws of physics are what they are, but these reasons may be partly historical and contingent, as in the case of biology. . . . [T]he desire to understand the world in terms of a naïve and radical atomism in which elementary particles carry forever fixed properties, independent of the history or shape of the universe, perpetuates a now archaic view of the world. It suggests a kind of nostalgia for the absolute point of view, a way of seeing the world that was lost when the Newtonian conception of space and time was overthrown.[37]

Now, can a recognition of, and dialogue with, other paradigms help Western science meet these challenges? The Pueblo science writer and educator Gregory Cajete suggests that, "with the creative influence of chaos theory and quantum physics, a new scientific cultural metaphor has begin to take hold" in the West. "The insights of this new science," he continues, "parallel the vision of the world long held in Indigenous spiritual traditions":

> Because of this undeniable parallel, Indigenous thought has the potential to inform a contemporary understanding of chaos. Such understanding allows modern consciousness to encompass the primal wisdom of Indigenous thought and with this to understand the fallacy of scientific and societal control. The modern obsession of being in control and the dream of eliminating uncertainty through control of nature, which is the underlying philosophical premise of Western science, must give way

> to the reality of moving creatively with the flow of events, which is the true reality of the universe.[38]

As we have seen, however, Western scientists are deeply divided in their response to such claims. Some, such as David Bohm, strongly echo Cajete's viewpoint:

> Understanding is now valued as the means to predict, control, and manipulate things. Of course, beginning with Francis Bacon, this has always been important but never so dominant as today. . . . This leads us to focus on particular problems, even when they are significantly related to a broader context. As a result, we fail to notice the unforeseen consequences, which cannot always be dealt with in terms of a fragmentary mode of thought. . . . For example, by exploring natural resources in a fragmentary manner, society has brought about the destruction of forests and agricultural lands, created deserts, and even threatens the melting of the ice caps. . . . Of course, many people think that solving these sorts of problems is only a matter of studying ecology or some other specialty. Certainly ecology does begin to acknowledge the complex dependence of each activity on the whole context. But really the problem is as much one of economics as it is of ecology, and this leads on to politics, and to the structure of society and the nature of human beings in general. The key issue is this: How is it possible to subject all these factors to prediction and control in order to manipulate the system and bring about good order? Clearly this is an impossible demand. . . . [H]ow can science lead human beings to control themselves? How do scientists propose to control hatred between nations, religions, and ideologies when science itself is fundamentally limited and controlled by these very things?[39]

Many others, though, enthusiastically endorse E.O. Wilson's *triumphalist* vision of reductionism, his conviction that, far from having been rendered 'naïve' or 'archaic', it now stands on the threshold of a glorious culmination:

> The legacy of the Enlightenment is the belief that entirely on our own we can know, and in knowing, understand; and in understanding, choose wisely. This self-confidence has risen with the exponential growth of scientific knowledge, which is being woven into an increasingly full explanatory web of cause-and-effect. . . . We now understand where humanity came from, and what it is. *Homo sapiens*, like the rest of life, was self-assembled. . . . Human autonomy having thus been recognized,

> we should now feel more disposed to reflect on where we wish to go. In such an endeavor it is not enough to say that history unfolds by processes too complex for reductionist analysis. That is the white flag of the secular intellectual, the lazy modernist equivalent of the Will of God. . . . We have begun to probe the foundations of human nature, revealing what people most need, and why.[40]

While Bohm and Wilson apparently stand worlds apart, Bohm here makes exactly the same mistake, hesitates on the same threshold, as Bohr: 'science', he claims, has been interested in the control and domination of nature "beginning with Francis Bacon". From the millennia-long perspective of Chinese or Native American science, this is an odd 'beginning' indeed. How, he asks, do "scientists" plan to "control" social tensions when "science itself" is socially "limited and controlled". Cajete and Leroy Little Bear might agree with this last claim: science is indeed *of*, not above, society, just as society is of, not above, the natural world. But, precisely because the science they are practicing *is* integrally related to a far greater, interdependent Whole, Native American scientists do *not* seek manipulative control over others: the whole philosophical and experiential basis of their methods and language of inquiry mitigates profoundly against such delusions of grandeur.

In the last years of his life, under the direct influence of the Native American paradigm, Bohm finally made this move into 'deep complementary': the leap from the 'world of science' to a world of *sciences*. The decisive factor seems to have been the radical difference in the *language* of inquiry in the two traditions.

For Bohr, as we have seen, the language of science, however esoteric or abstract, was ultimately rooted in the concepts and scales of everyday, commonplace communication, and *that* language, he assumed, was essentially the same in every cultural complex. In his 1980 book *Wholeness and the Implicate Order*, however, Bohm came to the drastic conclusion that 'science' could never evolve beyond a simplistic reductionism unless 'language' evolved beyond its own internal reductionism: the artificial grammatical separation of the world into 'subject' and 'object', 'verb' and 'noun.' Asking "whether there are any features of the commonly used language which tend to sustain and propagate" the endemic fragmentation of reality, Bohm comments:

> A cursory examination shows that a very important feature of this kind is the subject-verb-object structure of sentences. . . This structure implies that all action arises in a separate entity, the subject, and that, in cases described by a transitive verb, this action crosses over the space between them to another separate entity, the object. . . . This is a pervasive structure, leading in the whole of life to a function of thought tending to divide things into separate entities, such entities being conceived of as essentially fixed and static in their nature. When this view is carried to its limit, one arrives at the prevailing scientific worldview, in which everything is regarded as ultimately constituted out of a set of basic particles of fixed nature.[41]

In order to 'unfix' nature from this perspective—to "unpin" the "butterfly"—language itself would have to acquire *free movement.* This revolution, which Bohm proposed moving towards through an alternative language model he called the "rheomode" (from the Greek verb *rheo*, 'to flow'), would change "the syntax and grammatical form of language…so as to give a basic role to the verb rather than the noun." Why was this so important? Because "the verb describes actions and movements which flow into each other and merge, without sharp separations or breaks. Moreover, since movements are in general ways always themselves changing, they have in themselves no permanent pattern of fixed form with which separately existent things could be identified." Such a language, Bohm added, was needed to adequately express the "overall worldview" he was struggling towards, "in which movement is, in effect, taken as a primary notion, while apparently static and separately existent things are seen as relatively invariant states of continuing movement".[42]

Both before and after the release of *Wholeness and the Implicate Order*, Bohm and a small group of colleagues and supporters worked hard to develop the scheme beyond these broad, ambitious outlines. After a while, however, as David Peat remembers, "Bohm no longer pursued the rheomode": his "ideas . . . were intriguing, yet the response he got from most professional linguists . . . was discouraging; they found his ideas naïve and unconvincing." In addition to this cool reception, the rheomode group "began to use . . . verbs as stand-ins for nouns, defeating the very purpose for which the language [model] had been created. Transforming language was a project more difficult than Bohm had anticipated."[43]

Fortunately for the frustrated Western physicist, however, he soon encountered something else he hadn't anticipated—a set of living languages *brilliantly* reflecting and expressing the undivided flow of natural movement:

> In the last year of his life . . . Bohm met with a group of Native Americans who were all speakers of the strongly verb-based Algonquian family languages. (The participants included Blackfoot, MicMaq, Cheyenne, Ojibwa, and Soto.) Bohm was struck by their process-based vision of the world and by the way they themselves viewed the role played by their language. Here was a society, it seemed to him, that practiced what he had envisioned for his rheomode.[44]

If scientific concepts are necessarily, integrally connected to a broader language and deeper worldview, then *radically different* languages and worldviews will necessarily generate complementary *sciences*. The revelation both exhilarated and depressed Bohm and his wife:

> As the discussions continued, Bohm learned [more] about the process-based worldview . . . Everything is said to be in flux, and this constant change is reflected in their . . . language . . . Both Bohm and Saral were moved by the deeply spiritual outlook of the participants. Speaking about the arrival of the first Europeans, Bohm remarked, "It would have been better if we had never come."[45]

The participants themselves, however, seemed glad that *this* newcomer had moved so humbly across the border of scientific imperialism: "Bohm died a few weeks before the second of these circles, held in Banff, Alberta. When the . . . [group convened] on the first evening, it was discovered that, by chance, one chair had been left vacant. This was Bohm's chair, and several of the speakers addressed him directly as if his spirit were present."[46]

However impressed Bohm was by the Algonquin perspective, his next step would not have been to simply *adopt* the language, to somehow leap over his own tradition into a brave new worldview. He stood not on the brink of a *conversion* but rather a *dialogue* bound deeply to affect—in a creative, non-formulaic manner—his own approach.

How, more specifically, might such a dialogue proceed?

III. Complementary Healing? Chaos Theory and the Split Between 'Worlds'

The situation of many Native Americans in a world made 'modern' by the 'triumph' of the West is often described in terms of 'schizophrenia', a limbo-identity suspended between the irrevocable certainties of the past and the alien uncertainties of the future. Cajete, for instance, takes as the starting-point of his search for a renaissance of Indigenous science (and science education) the tragic premise that "American Indian people today live a dual existence. . . . They are constantly faced with living in a larger society that does not really understand nor respect their traditional life symbols, ecological perspectives, understanding of relationship to the land, and traditional ways of remembering to remember who they are. . . . The results for many Indian communities are 'existential' problems such as high rates of alcoholism, suicide, abuse of self and others, depression, and other social and spiritual ills."[47] As Cajete adds, however, such symptoms of profound dis-ease, and the *unthinkable* dislocations they reflect, are not limited to one, albeit acutely affected, section of the population:

> Tewa people call this state of schizophrenic-like existence *pingeh heh* (split thought or thinking, or doing things with only half of one's mind). As an Indian educator, I believe that modern Indian education ultimately has to be about healing this split. Healing the split is not a task for Indian people only. It is also the task of others who consider themselves people of place, and thereby experience alienation from mainstream society as do many Indian people.[48]

For David Bohm, prior to his transforming encounter with Native America, the scientific obsession with the control and domination *of* nature—a compulsion normalized, both reaffirmed and disguised, by the basic structures of language—exemplified a broader 'fragmentation of thought' artificially separating humanity *from* nature. Given both the scale of the resultant crisis, and the growing failure of the reductionist paradigm to provide a coherent conceptual framework for inquiry, Bohm argued that we "need to change what we mean by 'science'. The moment has come for a creative surge along new lines."[49]

For Bohm, emerging Western ideas about change and transformation in natural systems—popularly referred to as 'chaos theory'—provided an ideal opportunity to open the scientific mind to the possibilities of such a redirection. For Cajete, these new shoots lent themselves naturally to comparison, mutually-affirmative rather than judgmental, with deep-rooted Native American perspectives:

> Chaos theory, derived from the cutting edge of Western scientific research itself, implies that systems are beyond the ability of scientists to predict or control except at the most superficial levels, and that all of nature is a chaotic system. Rather than seeking to control natural reality, Native science focuses its attention upon subtle, inner natures wherein lie the rich textures and nuances of life. This is exactly what chaos theory shows us: small, apparently insignificant things play major roles in the way a process unfolds. Indeed, Native science may be said to be the 'science of the subtle'.[50]

Interestingly, the subtitle of the classic popular presentation of the new Western ideas—*Order Out of Chaos*, by Ilya Prigogine, Nobel Laureate for Chemistry, and his colleague Isabelle Stengers—is *Man's New Dialogue With Nature*. Why is such a dialogue necessary? To heal what the great scholar of Chinese science Joseph Needham described as the "characteristic European schizophrenia",[51] the false split between the human and the natural worlds which has proved so damaging, and may yet prove fatal, to 'both':

> In classical science the emphasis was on time-independent laws. . . . It is natural that this quest for an eternal truth behind changing phenomena aroused enthusiasm. But it also came as a shock that nature described in this way was in fact debased: by the very success of science, nature was shown to be an automaton, a robot. . . . Modern science transmuted this . . . stance into what seemed to be an established truth; and this truth, the reduction of nature to atoms and void, in turn gave rise to what Lenoble has called the "anxiety of modern men". How can we recognize ourselves in the random world of the atoms? Must science be defined in terms of rupture between man and nature?[52]

Yes, the authors answer, *if* this scientific worldview, however tragic, is based on an accurate reading and representation of fundamental natural processes. The basic, twofold conviction of pre-complementarity Western science was that: 1) immutable, eternal laws exist behind the phenomenal veil; and 2) that the quantitative, reductionist approach—analytically dissecting (*purposefully* rupturing) the non-living "machinery of life"—was capable of piercing the veil, reading the secret thoughts of that 'godlike mind'. From the seventeenth to the twentieth century, faith in Western science (understood, of course, as synonymous with 'science itself') meant faith in the *revelatory* quality of this basic procedure.

Reductionism, in this sense, played the part of *resurrection* in Christian faith. As St. Paul observed (1 Corinthians, 13: 19), "if Christ has not been raised, then our preaching is in vain and your faith is in vain. We are even found to be misrepresenting God, because we testified of God that he raised Christ . . . If for this life only we have hoped in Christ, we are of all men most to be pitied." An 'Age of Reason' version of the text might read: "if reductionism does not reveal, lift us above nature, then our experiments and theories are in vain. We are even found to be misrepresenting nature, because we prophesied that it would reveal its secrets at our hands . . . If we are learning only how to *speak* about nature, our delusion is surely most to be pitied."

Prigogine and Stengers' point, of course, is that this pity can often be extended to those who believe reductionism *does* fulfil a revelatory function: the revelation, for many devotees, is of a meaningless, alien Universe, and thus ironically *unsalvational* in nature. As one of the major new testaments to such despair, *Chance and Necessity* by Jacques Monod (1972), reads: "Man must at last . . . awake to his total solitude, his fundamental isolation. Now does he realize that, like a gypsy, he lives on the boundary of an alien world. A world that is deaf to his music, just as indifferent to his hopes as it is to his suffering or his crimes." Prigogine and Stengers' respond: "This is a paradox. A brilliant breakthrough in molecular biology, the deciphering of the genetic code, in which Monod actively participated, ends upon a tragic note. This very progress, we are told, makes us the gypsies of the universe. How can we explain this situation? Is not science a way of communication, a dialogue with nature?"[53]

Chaos theory, they then argue, suggests that the "tragic choice"[54] posed by Monod—to accept the reality of the disenchanted cosmos or believe in the fairytales of humanity's childhood—is an entirely false construction. A fresh channel of communication with nature is indeed in the process of being opened in the West by, as Cajete put it, "the cutting edge" of research exposing the Enlightenment paradigm as "an excessive simplification" compared by Prigogine and Stengers to "reducing buildings to piles of bricks":

> Yet out of the same bricks we may construct a factory, a palace, or a cathedral. It is on the level of *the building as a whole* that we apprehend it as a creature of time, as a product of culture, a society, a style. But there is the additional and obvious problem that, since there is no one to build nature, we must give to its very 'bricks'—that is, to its microscopic activity—a description that accounts for this building process.[55]

What account, then, does chaos theory give of this process, and to what extent is its "new dialogue" resonant with the frequencies of human-natural communication long open in the Native American tradition? Let us juxtapose our two main sources:

> Chaos is both movement and evolution. It is the process through which everything in the universe becomes manifest and then returns to the chaos field. The flux, or ebb and flow, of chaos appears in everything and envelops us at all times and in all places. From the evolving universe to the mountain to the human brain, chaos is the field from which all things come into being. No wonder Native science envisions the spirit of the natural world alive with disorder becoming order and all the mystery of mirrored relationships. . . . Self-organization or 'creativity' out of chaos occurs everywhere in nature. Random interstellar gases and electromagnetic fields of radiation self-organize to form galaxies and star systems. The interaction of rain with the earth's geological landscape leads to the vast patterns of rivers and streams that form drainage systems. Birds or insects fly in perfect unison.[56]

> Our universe has a pluralistic, complex character. Structures may disappear, but also they may appear. . . . Today we see everywhere the role of irreversible processes, of fluctuations. The models considered by classical physics seem to us to occur only in limiting situations such as we can create artificially by putting matter into a box and then waiting till it reaches equilibrium. The artificial may be deterministic and reversible. The natural contains essential elements of randomness and irreversibility. This leads to a new view . . . in which matter is no longer the passive substance described in the mechanistic worldview but is associated with spontaneous activity. This change is so profound that . . . we can really speak about a new dialogue of man with nature.[57]

And again:

> Then there is the notion of subtle influences, or the 'butterfly effect' in chaos theory. In chaotic systems, even small things turn out to have large-scale effects over a period of time. For example, if we look at

> weather we see a recurring climatic pattern over a long period of time. However, if we examine details we see that weather is in constant flux due to the bifurcating and amplifying activity of a host of subtle effects. In a weather system, everything is interconnected. Positive and negative feedback loops are in constant motion, and somewhere in the system, a 'butterfly' loop may cause slight changes. Sooner or later, one of these loops is amplified, and we see a dramatic and unpredictable shift in the pattern. The butterfly effect may be called chance, but it is really the cumulative influence of a small change in a system. It may be an increase or decrease of temperature in a weather pattern, an individual such as Gandhi taking a stand against oppression, or a Native prayer, song, dance, or ritual to bring rain to a parched land. In the world of chaos, anything is possible.[58]

> . . . from this perspective life no longer appears to oppose the 'normal' laws of physics, struggling against them to avoid its normal fate—its destruction. On the contrary, life seems to express in a specific way the very conditions in which our biosphere is embedded, incorporating the nonlinearities of chemical reactions and far-from-equilibrium conditions imposed on the biosphere by solar radiation. We have discussed the concepts that allow us to describe the formation of dissipative structures, such as the theory of bifurcations. It is remarkable that near-bifurcations systems present large fluctuations. Such systems seem to 'hesitate' among various possible directions of evolutions . . . A small fluctuation may start an entirely new evolution that will drastically change the whole behavior of the macroscopic system. The analogy with social phenomena, even with history, is inescapable. Far from opposing 'chance' and 'necessity,' we now see both aspects as essential in the description of non-linear systems far from equilibrium.[59]

Cajete would certainly agree with Prigogine and Stengers that "this perspective"—that the "universe has a pluralistic . . . character"—creates a reputable scientific basis for dialogue with nature. Would Prigogine and Stengers, however, agree with Cajete that such a dialogue should also open channels of communication between different traditions, perspectives reflecting the pluralistic character of the *scientific* universe? *Order Out of Chaos* suggests that they stand alongside Bohr on the *brink* of such deep complementarity:

> We believe that we are heading toward a new synthesis, a new naturalism. Perhaps we will be able to eventually combine the Western tradition, with its emphasis on experimentation and quantitative formulations,

> with a tradition such as the Chinese one, with its view of a spontaneous, self-organizing world. Toward the beginning of this Introduction, we cited Jacques Monod. His conclusion was: "The ancient alliance has been destroyed; man knows at last he is alone in the universe's indifferent immensity out of which he emerged only by chance." Perhaps [in one sense] Monod is right. The ancient alliance has been shattered. Our role is not to lament the past. It is to try to discover in the midst of the extraordinary diversity of the sciences some unifying thread.[60]

By the "ancient alliance", Monod designated the profound, participatory connection between humanity and nature: Life as Sacred, Earth as Mother, Cosmos as Home. A belief in *that* natural order, however, is *still* the "unifying thread" of traditions "such as the Chinese", and most definitively including the Native American. But note that Prigogine and Stengers do not seek such a thread in the midst of a plurality of *sciences* but rather in the diversity of *the* sciences: that is, the modern Western triad of physics, chemistry and biology. *That* tradition, they believe, is capable of evolving, developing a more natural view of nature, by incorporating new conceptual elements from outside. But outside what: the West, or *science*? If only one truly scientific framework is assumed—the triad as the retort in which this 'synthesis' occurs—is not the process actually *assimilative*, an artificial transmutation of integral, *wholly different* traditions, 'entities' rather than 'elements', separate rather than separable? And if there is *more* than one framework—a plurality of receptacles shaping knowledge of nature—why is synthesis rather than dialogue, integration rather than collaboration, desirable or necessary?

If Prigogine and Stengers are right, then the Western tradition clearly stands in need of a major conceptual and philosophical overhaul, a process it may well have to look outside of its own worldview to complete. But does it follow from this that traditions already possessing a strong and coherent conceptual framework have paid for it by neglecting the alleged strengths of the "Western tradition," the "emphasis" on "experimentation" and "quantitative formulations"? As Peat seeks to demonstrate, it is a Western *myth* that the Native American tradition does not conduct an intense empirical experimental relationship with the natural world. The context and content, the regulation and organization, of that relationship is not inferior but *complementary* to the West's, shaped by a different paradigm of inquiry. As Leroy Little Bear points out, from the perspective of *his* worldview, Western science seems burdened and distracted by an *over*emphasis on mechanistic and quantitative formulations, blinded to other, perhaps far more significant, organic and qualitative, factors of *real* change and transformation. Does this worldview, or the Chinese, require *synthesis* with the West's? Or is it, on the contrary, sufficiently self-assured and independent to welcome an exchange of information and opinion?

The discussion leads back to the insight at the heart of complementarity: the inherent partiality—the essential 'unsynthesizability'—of the 'truths' disclosed by science. As Peat stresses, the issue is not which scientific tradition is *right* or *better*; and the objective should not be to combine the best of all paradigms into a mega- or super-paradigm. The results of any such 'grand unified theory' will necessarily be *synthetic*: artificial and, by a supreme irony, unnatural. No science can hope to grasp, i.e. demystify, the *wholeness* manifested in the objects of its inquiry; no science can hope to rise above the cultural environment both shaping and shaped by its activity: the existence in diverse environments of distinct understandings of natural phenomena *is itself a natural phenomenon*. For Prigogine and Stengers to acknowledge the limits, even the unnaturalness, of the traditional Western framework is certainly to take a step away from the exclusionary supremacism voiced by Wilson. To seek to overcome this weakness, however, through pursuit of a synthetic Grail, is to step *back* from the brink of genuine pluralism: to redraw, albeit more expansively, the line separating *the* circle of science from other scientific worlds.

For the Western circle to really open, a very different step will have to be taken, one far more modest and reasonable than any "new synthesis", yet requiring the rehabilitation, *within the Enlightenment tradition*, of a demon long thought exorcised: metaphysics.

IV. Conclusion

Escaping the Vice

Prigogine and Stengers ask: "Must we choose between a science that leads to alienation and an anti-scientific metaphysical view of nature?" Not so, they insist: "We think such a choice is no longer necessary, since the changes that science is undergoing today leads to a radically new situation. . . . Modern science originated in the specific context of the European seventeenth century. We are now approaching the end of the twentieth century, and it seems that some more *universal message* is carried by science, a message that concerns the interaction of man and nature as well as of man with man."[61] Such a message, they add, will reflect "the emergence of new conceptual structures that now appear as essential to our understanding of the physical world—the world that includes us."[62]

"Our vision of nature is undergoing a radical change":[63] but *whose* vision, exactly? Presumably, the 'children' of the Age of Reason, the inheritors of the picture of the world bequeathed, and framed, by the "specific context of the European seventeenth century". But here, again, we both reach the threshold of a true breakthrough and fall back on false distinctions. For where other than Enlightenment Europe was the "metaphysical view of nature" first identified and denounced as "anti-scientific"? The self-image of modern Western science is created in the heroic empirical gaze of Galileo, refusing to view natural reality through the dogmatic prism of scholastic theology. The self-esteem of the entire Age of Reason is inseparable from the pride taken in this historic break of physics from metaphysics, science from superstition, chemistry from alchemy, etc. It was in this particular *atmosphere*—as the product of a "culture, a society, a style"—that metaphysics came to be regarded as *synonymous* with theology, an approach neither complementary nor supplementary but simply *antithetical* to the search for truth, incommensurate with an accurate and reliable understanding of nature.

Now, the phrase used by Prigogine and Stengers—"an anti-scientific metaphysical view of nature"—suggests the possibility of a legitimate metaphysics *of* scientific inquiry. But nowhere in *Order Out of Chaos* is such a possibility explored. In fact, the authors explicitly advocate a "third path" between the "positivistic" and the "metaphysical", namely to "ask whether the simplicity" of the processes of change and transformation "traditionally considered in physics and chemistry" was "due to the fact that attention was paid mainly to some very simplified situations, to heaps of bricks in contrast with the cathedral to which we have alluded."[64]

This proposed solution is very much in the spirit of Bohr: to avoid conceptual confusion and theoretic delusions about nature, science needs to pay more attention *to the ways in which it pays attention* to nature, needs to appreciate the ways those ways radically limit the scope of the claims it can make. Indeed, it is only by means of such a retreat, the contraction of the applicability of classical concepts, that advances in understanding are made possible. "Curiously," as Prigogine and Stengers say, "the unexpected complexity that has been discovered in nature has not led to a slowdown in the progress of science, but on the contrary" to a radically new picture of natural reality.

All of which may be true. But the question before us is this: can *any* picture of physical reality be drawn without a frame-of-reference—a set of assumptions about the *nature* of that reality—guiding the inquiring hand? In asking us to think about the way we think about nature, is not Bohr, is not Prigogine, asking us to examine just such a set of assumptions? And what other than meta-physical can such a scientific *setting* be?

Reviewing the legendary debate between Bohr and Einstein over quantum theory, Robert Pirsig notes it "was always" shaped "in terms of thought experiments", logic-games potentially illuminative *of*, but in this case substituted *for*, deeper questions: "Although Bohr had said, 'Reality is a term we must learn to use,' the debate was never raised to the level of a discussion of what this 'physical reality' is whose description is either complete or incomplete." "The reason" for this lacuna, or blindspot, Pirsig argues, was that "in those days a philosophic discussion of 'reality' was discouraged":

> Discussions of reality were metaphysics and metaphysics was something associated with medieval religious mysticism. Yet as I read through the material . . . I could see that this was not primarily a quarrel about physics, it was about metaphysics. . . . There is no way one can possibly construct a scientific experiment to determine whether or not an external reality exists if there is a difference in metaphysical interpretation. Whatever results you come up with can still be explained differently in each metaphysical system.[65]

Pirsig's conclusion? It "is necessary to get into a closer look at the metaphysical system of Complementarity itself"[66]—exactly the step Bohr's (and Einstein's) own modern scientific prejudice prevented him from taking. Both Bohr and Einstein believed desperately in the importance of a philosophical dimension to science, yet neither were able, or at least *culturally disposed*, to see the necessarily metaphysical dimensions of *any* such philosophical perspective.

As Prigogine and Stengers make clear, 'those days' are still with us. And this persistent prejudice, I believe, lies at the root of another, the obstacle blocking the transition from shallow to deep complementarity: the prejudice of Western science against other sciences. Different scientific 'worlds', complementary styles and modes of inquiry, are possible *only* because different metaphysical *atmospheres* exist to sustain them. But not only does the Western paradigm construct a 'one-world' model of science; it refuses to acknowledge *that* even that world, its own, is a reflection and expression of an underlying metaphysics. And when a metaphysics, instead of providing a natural conceptual atmosphere of inquiry, is ignored—pervasively present but invisibly dispersed in the 'pure' air of 'science itself'—then the 'method' that science supposedly just 'is' runs the risk of becoming a *madness*, the classic delusion Bohr was so keen to dispel: a failure to think straight, valorized and experienced as straightforward thinking.

Both quantum and chaos theory clearly pose awkward and deeply unsettling questions for the Western scientific paradigm. Bohr and Prigogine, however, seek to treat that dis-ease without recourse to metaphysical examination: *without questioning* the Enlightenment definition of 'science' as a set of rational methods of inquiry, a yardstick against which other approaches should be measured.

In the Chinese and Native American traditions, by contrast, scientific understandings of physical reality are *explicitly* located within a broader and deeper understanding of *the* reality—the Way, the Great Spirit—manifested in natural phenomena. Starting from the premise that a metaphysical frame-of-reference necessarily obscures a true picture of reality, what conclusion can you draw from such traditions other than Wilson's—that they are *un*scientific—or Bohr's—that they are complementary *to* science? Unless Western science acknowledges the metaphysical basis of its own endeavors, it will prove unable to resolve the essentially *non-physical* issues encoded in the Bohr-Einstein dispute. And until Western science takes that step, back across the metaphysical shadow it claims not to cast, it will remain blind to the validity of other sciences, other ways of partially knowing the true Way.

As stressed above, the 'madness' of anti-metaphysical science is symptomatic, even emblematic, of a wider disorder, the general fragmentation of modern life and thought; the dis-integrative, and doomed, attempt to *control life through thought*. In the introduction to *Process and Reality*, his momentous attempt to restore metaphysical vision to Western science, Alfred North Whitehead argued that the prime task of philosophy now lay in "the welding of imagination and common sense into a restraint upon specialists, and also into an enlargement of their imaginations. By providing the generic notions philosophy should make it easier to conceive the infinite variety of specific instances which rest unrealized in the womb of nature."[67] Under the influence—in the expansive metaphysical atmosphere—of these 'generic notions,' the claims made on behalf of specialized inquiry contract, located in an overall appreciation of the Whole. If, however, specialization provides both picture and frame, the Whole is inevitably, *systematically*, obscured, reduced to partial, separable 'components', unnaturally torn from the 'womb':

> We habitually speak of stones, and planets, and animals, as though each individual thing could exist, even for a passing moment, in separation from an environment which is in truth a necessary factor in its own nature. Such an abstraction is a necessity of thought, and the requisite background of systematic environment can be presupposed. That is true. But it also follows that, in the absence of some understanding of the final nature of things, and thus of the sorts of backgrounds presupposed in such abstract statements, all science suffers from the vice that it may be combining various propositions that which tacitly presuppose inconsistent backgrounds. No science can be more secure that the unconscious metaphysics which tacitly it presupposes. The individual thing is necessarily a modification of its environment, and cannot be understood in disjunction. All reasoning, apart from some metaphysical reference, is vicious.[68]

Whitehead was writing shortly before the 'birth' of the atomic age, yet his sense of the possible consequences of unrestrained fragmentation in the human view of nature seemed already to encompass the apocalyptic. "Systems, scientific and philosophic, come and go," he argued, with the "transitions" from "exhausted" methods of "limited understanding" to "new fruitfulness" of insight achieved "by recurrence to the utmost depths of intuition for the refreshment of imagination." But, precisely because such transitions are creative, unpredictable in essence, their success can never be guaranteed: "opportunity" for change "leads upwards or downwards." Clarity, for example, is required on the purpose of the transition: what does it provide an opportunity *for*? In Whitehead's view, *the reestablishment of dynamic relations between physics and metaphysics*. In the modern age of mass destruction, already glimpsed in the trenches and poison clouds of World War I, failure to achieve this rapprochement could prove catastrophic. "Philosophy," he urged, "should now perform its final service" to humanity, no less than to "seek the insight, dim though it be, to escape the wide wreckage" of "race" and planet.[69]

Decades after the likely form of this 'wreckage' appeared in the mushroom cloud, the physicist-turned-philosopher Carl-Friedrich von Weiszäcker wrote:

> Physics, as it is practiced in modern times, characteristically does not really ask what matter is, biology does not ask what life is, and psychology does not ask what the soul is; instead, these terms just vaguely circumscribe the area one intends to investigate. . . . As a result, science, which leaves the fundamental questions aside, has progressed incredibly fast in comparison with the very slow . . . process of philosophical reflection, which truly confronts these difficult questions. On the other hand, we must not deceive ourselves: *the methodological procedure of science just characterized has something murderous in it if it no longer knows how questionable it is*. Those questions are difficult, but they are not unimportant. Heidegger's formula 'Science does not think' can hardly be quoted to any scientist without provoking anger. In Heidegger's sense of the word 'think,' however, the formula is literally correct. For Heidegger takes 'to think' as meaning 'to put oneself in question once more,' and precisely this science will not do in its normal practice. But it must be done if science should someday be able to relate to the living human being, who is a partner in life and not merely an object.[70]

For both Whitehead and von Weiszäcker, this 'viciousness'—the more than somewhat 'murderous' quality and potential of 'unthinking' modern science—should be considered not as a tendency inherent to 'science itself" but rather a consequence of the artificial impoverishment of scientific identity, a contraction compensated by an *overinflation* of ambition and claim. As Whitehead concludes, "the Certainties of Science are a delusion.

They are hedged around with unexplored limitations. Our handling of scientific doctrines is controlled by the diffused metaphysical concepts of our epoch."[71]

The point was made in full awareness of the then freshly-minted 'Copenhagen Interpretation' of quantum physics, built around the uncertainty principle and complementarity; his critique is not directed at the outmoded certainties of the classical, Newtonian age. But Western science, he maintained, remained classically deluded in its certainty that *metaphysics* was outmoded, superceded by its rational mode of inquiry—a view of the world, however provisional, obtained without aid of a *worldview*.

After so long, can this 'diffusion' still be reversed, metaphysical clarity lent to the incipient Western quest for a holistic perspective? The answer, I believe, is that the *potential* for such a transition exists, the path laid open by the 'cutting-edge' of Western research, and that its pursuit would clear the way for resonant, complementary relations with other scientific traditions. Indeed, establishing this dialogue is likely to *accompany* rather than follow the transition, helping guide it 'upward' to 'new fruitfulness'.

For the opportunity to be taken, however, Western science—and society more generally—has first to revise its opinion of these other traditions as less than scientific. Only by embracing such 'deep complementarity', relinquishing its monopolistic hold on 'science', can it hope to recognize itself as *more* than scientific: part of a far greater cultural, natural, human whole.

A butterfly that saw itself as *of*, not *above*, the forest, would be a very different creature.

References

1. This paper was inspired and deeply informed by my experience in leading an interdisciplinary seminar group on 'Integrative Science and Philosophy' at Cape Breton University. I wish to record my indebtedness to all members of the group, both student and faculty, for their participation and insights. Special thanks, for support and funding, are due to Canada Research Chair in Integrative Science, Dr. Cheryl Bartlett, Director of the University's Integrative Science Program (Toqwa'tu'kl Kjiijitaqnn).

2. Henri Bergson, *Creative Evolution* (New York: Barnes & Noble Books, 2005), p. 59.

3. Jeremy W. Hayward, *Letters to Vanessa: On Love, Science, and Awareness in an Enchanted World* (Boston: Shambhala, 1997), viii.

4. Niels Bohr, *Atomic Physics and Human Knowledge* (New York: John Wiley & Sons, 1957), p. 2.

5. Quoted in Nick *Herbert*, *Quantum Reality* (New York: Anchor Press/Doubleday, 1985), p. 259.

6. Bohr, *Atomic Physics and Human Knowledge*, p. 2.

7. Bohr, *op.cit.*, p. 20.

8. Bohr, *op. cit.*, p. 79.

9. F. David Peat, *From Certainty to Uncertainty* (Washington, D.C.: Joseph Henry Press, 2002), p. 8.

10. Werner Heisenberg, *Physics and Philosophy* (London: Penguin Books, 1990), p. 194; first published by Harper & Row, New York, 1962.

11. Quoted in Helge Krage, *Quantum Generations: A History of Physics in the Twentieth Century* (Princeton, NJ: Princeton University Press, 1999), p. 172.

12. Krage, *op. cit.,* p. 172.

13. Robert M. Pirsig, 'Subjects, Objects, Data and Values,' in *Einstein Meets Magritte: an Interdisciplinary Reflection on Science, Nature, Art, Human Action and Society,* Volume 1 (The White Book) (Brussels: VUB University Press, 1999), p. 79-98.

14. Stephen W. Hawking, *A Brief History of Time* (London: Bantam Press, 1988), p. 175.

15. David Bohm & F. David Peat, *Science, Order & Creativity* (London: Routledge, 1987), p. 81.

16. Bohr, *op. cit.,* p. 84.

17. Heisenberg, *op. cit.*, p. 166.

18. *Op. cit.*, p. 166.

19. Fritjof Capra, *The Tao of Physics,* Third edn. (London: Flamingo, 1992), p 175.

20. Capra, *op. cit.*, pp. 174-175.

21. *Op. cit.*, p. 175.

22. Heisenberg, *Physics and Beyond: Encounters and Conversations* (San Francisco: Harper Torchbooks, 1972), p. 63.

23. F. David Peat, *Blackfoot Physics: A Journey Into the Native American Universe* (Grand Rapids, MI: Phanes Press, 2002), pp. xi-xii.

24. Edward O. Wilson, *Consilience: The Unity of Knowledge* (London: Little, Brown and Company, 1988).

25. F. David Peat, *op. cit.,* p. xiv.
26. Peat, *op. cit.*, p. 130.
27. *Op. cit.*, p. 130.
28. *Op. cit.,* p. 131.
29. *Op. cit.*, p. xiv.
30. Bohr, *op. cit.*, p. 9.
31. Peat, *op. cit.*, p. 241.
32. *Op. cit.*, p. 241.
33. Leroy Little Bear, Foreword to Gregory Cajete, *Native Science: Natural Laws of Interdependence* (Santa Fe, NM: Clear Light Publishers, 2000), pp. ix-x.
34. Leroy Little Bear, *op. cit.*, p. x.
35. Ian Stewart, *Nature's Numbers: the Unreal Reality of Mathematics* (New York: Basic Books, 1995), p. 147.
36. Leroy Little Bear, *op. cit.*, p. xi.
37. Lee Smolin, *The Life of the Cosmos* (London: Weidenfeld & Nicolson, 1997), p. 18.
38. Gregory Cajete, *Native Science, op. cit.,* p. 16.
39. David Bohm & F. David Peat, *Science, Order & Creativity,* (London: Routledge, 1987), pp. 11-13.
40. Wilson, *Consilience, op. cit,* pp. 331-332.
41. David Bohm, *Wholeness and the Implicate Order* (London: Ark Paperbacks, 1983), pp. 28-29.
42. Bohm, *op. cit.*, pp. 29-30.
43. F. David Peat, *Infinite Potential: The Life and Times of David Bohm* (Reading, MS: Addison Wesley, 1997), p. 251.
44. Peat, *op. cit.*, pp.251-252.
45. *Op. cit.*, p. 315-316.
46. *Op. cit.*, p. 316.
47. Gregory Cajete, '"Look to the Mountain: Reflections on Indigenous Ecology,' in Cajete, Ed., *A People's Ecology: Explorations in Sustainable Living* (Santa Fe, NM: Clear Light Publishers, 1999), pp. 16-17.
48. Cajete, *op. cit.,* p. 17.
49. David Bohm & F. David Peat, *Science, Order & Creativity* (London: Routledge, 1987), pp. 11.
50. Cajete, *Native Science, op. cit.,* pp. 16-17.
51. Quoted in Ilya Prigogine & Isabelle Stengers, *Order Out of Chaos: Man's New Dialogue With Nature* (New York: Bantam Books, 1984), p. 7.
52. Prigogine & Stengers, *op. cit.*, p. 3.
53. *Op. cit.*, p. 3.

54. *Op. cit.*, p. 7.
55. *Op. cit.*, p. 7. Emphasis added.
56. Cajete, *Native Science, op. cit.*, p. 16.
57. Prigogine & Stengers, *op. cit.*, p. 9.
58. Cajete, *op. cit.*, p. 18.
59. Prigogine & Stengers, *op. cit.*, p. 14.
60. *Op. cit.*, p. 22.
61. *Op. cit.*, p. 7.
62. *Op. cit.*, p. 2.
63. *Op. cit.*, p. 2.
64. *Op. cit.*, pp. 10-11.
65. Robert M. Pirsig, 'Subjects, Objects, Data and values,' *op. cit.*, p. 84.
66. Pirsig, *op. cit.*, p. 85.
67. Alfred North Whitehead, *Process and Reality*, originally published 1928, Corrected edn. (New York: The Free Press 1978), p. 17.
68. Whitehead, *Adventures of Ideas*, originally published 1933 (New York: The Free Press, 1967), p. 154.
69. Whitehead, *op. cit.*, p. 159.
70. Carl Friedrich von Weiszäcker, *The Unity of Nature*, (New York: Farrar, Straus and Giroux, 1980), p. 233. Emphasis added.
71. Whitehead, *op. cit.*, p. 154.

Edy Korthals Altes

Edy Korthals Altes *was born in the Netherlands in 1924 and joined the Foreign Service in 1951. After completing his studies at the Rotterdam School of Economics, he served in New York, Colombo, Paris, The Hague, Bonn, Rome, Djakarta, Brussels and was Ambassador to Poland and Spain. In July 1986 he was granted an honourable discharge following his voluntary resignation when he took a public stand on the arms race (in particular Star Wars). Since then he has been active in the Pugwash movement for peace and security. He has served faith-based organisations both in the Netherlands and elsewhere in Europe. He has also acted as Chairman of the International Affairs section of the Netherlands Council of Churches.*
From 1994-1999 he was President of the World Conference of Religions for Peace (WCRP) and is a member of SPES (Spirituality in Economics and Society) at the University of Leuven. His books include Heart and Soul for Europe: An essay on spiritual renewal, *Assen, Van Gorcum, 1999.*

THE CONTRIBUTION OF RELIGIONS TO A JUST AND SUSTAINABLE ECONOMIC DEVELOPMENT

EDY KORTHALS ALTES

The road towards a just and sustainable economic development is long and arduous. Will it be possible to reach the goal in time? Yes! says Jeffrey Sachs, the well- known UN expert on poverty: "the end of poverty is within reach of modern technology and our financial possibilities." Yes! claim Donella and Dennis Meadows and Jorgen Randers in the latest report on the "Limits to Growth".[1] Yes! declared the government leaders of more than 180 countries in September 2000 when they signed the "Declaration on the Millennium Development Goals". In this important Declaration they solemnly committed their countries to halve, by the year 2015, the proportion of the world's people whose income is less than one dollar a day and the proportion of people who suffer from hunger and, by the same date, to halve the proportion of people who are unable to reach or to afford safe drinking water.

All very positive intentions indeed; it would however not be the first time that high-sounding words were not realized!

The actual discussion on sustainable development is mostly oriented towards pragmatic solutions. High expectations exist regarding the possibility to solve the environmental crisis through a combination of technological inventions, fiscal incentives and other policy measures. Certainly, all of these are important but more will be needed. A just and sustainable development will not be achieved without a fundamental change in attitude towards human beings, goods and nature. A change based on a renewal of spirit. Religions could be a decisive factor to bring about this decisive change.

The contribution of religions to a just and sustainable development

There are compelling reasons for religions to be seriously involved in the debate on a new—more promising—approach in economics. The present course of economic development—particularly in the highly developed economies—is bound to lead to a collapse. Religions, faithful to their prophetic calling, should stand up whenever "the humane" is threatened or the integrity of creation trampled upon. This is a responsibility for all religions, which should receive high priority. With their vast numbers of adherents spread all over the world, religions have a great potential to mobilize resources to assure a truly humane existence for all people on earth. But the road towards a sustainable and just society is difficult to find and reliable road maps rare. Neither the Bible, the Qur'an nor any other sacred Scripture holds a recipe for a new economic order. Still, they can provide us with reliable guidelines about the direction to follow.

A detailed analysis of the relation between the basic teachings of various religions with present economic thinking and acting would therefore be of considerable interest. Within the context of this essay this is unfortunately not possible. As Christian churches have been particularly active since the end of the 19th century in formulating a critical stand on social and economic affairs I will first give a brief sketch of their positions followed by a few observations about important developments in Buddhism and Sufism. Another—often overlooked—reason for the emphasis given to prophetic voices within Christianity is the sad fact that the present life-threatening economic developments are chiefly the result of policies and practices in modern nations wherein Christians form a majority!

Elements in biblical thinking

Of fundamental importance in biblical thinking is the awareness that we are living in God's world. At first sight this may seem a ludicrous thought for modern human beings so infatuated with an inordinate sense of self-importance. But would it not be a good thing—in a culture in which systematic doubt prevails—to ask us whether it is not a colossal mistake to push the basic notion of Transcendence beyond our horizon? A sensible question, so it seems to me, in a civilization in dire need for sources of inspiration.

Biblical thinking challenges directly the large scale process of destruction and pollution of our natural environment that is now taking place in the name of "economics". Certainly, in Genesis 2 we read that God said to man . . ."and subdue the earth—mastering the fish in the sea, the birds of the air and every living creature that crawls on earth . . ." But this much-abused text in the ecological debate does not mean that human beings are entitled to do whatever they like with the earth and its creatures. These words are not addressed to so-called "autonomous beings" but to men and women called upon to love God with their whole being. The earth is, after all, entrusted to human beings who are fully responsible to God and not to an abstract *homo oeconomicus*! Listening with new ears to the Principal Commandment would bring us to acknowledge the enormity of the mistake of the maltreatment of our planet in the name of a false concept of economics. The Bible provides no justification whatsoever for the present irresponsible human behaviour towards nature. This is why all those who are professing faith in God should become far more active in protesting against the destruction of our natural habitat.

Prophetic voices

A document of great importance was the Papal Encyclical *Rerum Novarum* (1891) on "new matters of misery, injustice and exploitation created by industrialization". Since then the Roman Catholic Church has consistently protested against the subordination of human well-being to economic ends. Time and again the social encyclicals affirmed the priority of the spiritual and ethical aspects in life.

Also within Protestantism there has been an impressive development of ecumenical social thought. A beginning was made in 1925 in Stockholm, during the world conference of the Life and Work movement. Another important moment was the formation of the World Council of Churches

(WCC) in Amsterdam in 1948. The ecumenical movement has since then wrestled with the role of the churches in a rapidly changing world. The persistence of poverty and the threats to the earth's capacity to sustain human life became ever more clearly recognized.

During the Vancouver Assembly of the WCC in 1990, the appeal was launched for a worldwide commitment on the issues of justice, peace and integrity of creation. This was not just another proclamation but seen as part of the essence of being a living church. Considerable reflection has been going on in this area. In present efforts to develop a theology of life much attention is given to remind society of the basic truth that all exercise of power—including economic power—is accountable to God. Prominence is also given to:

- The Biblical option for the poor
- The equal value of all races and peoples
- The creation as beloved of God
- The peace of Jesus Christ.

At the same time another significant development took place within the World Alliance of Reformed Churches (WARC). During its general council meeting in Debrecen, in the summer of 1997 it declared, "the struggle against economic injustice and ecological destruction is at the very center of Christian faith". The 211 member churches were called upon to "work towards a confession of their beliefs about economic life which would express justice . . . reflect priority for the poor and support an ecologically sustainable future".

This line of thinking finally resulted in the Accra Confession (August 2004), signed by the representatives of all the participating churches, representing more than 100 million Christians. An important achievement after the intense discussions between the Northern and Southern churches. In this significant document an analysis is given of the appalling consequences of the present economic system for countless people and . . . nature. It points to the dramatic convergence between the suffering of the people and the damage done to creation. In doing so it plainly rejects "the culture of rampant consumerism and the competitive greed and selfishness of the neo-liberal global market system, or any other system, which claims there is no alternative". The declaration strongly affirms that: "the integrity of faith is at stake if we remain silent or refuse to act."

Another initiative of immediate relevance for the struggle against poverty is the recent worldwide campaign of the Anglican Church. Its strong advocacy for debt relief for the poor countries could significantly contribute to alleviate the misery of millions of people.

It goes without saying that this growing awareness about the nefarious consequences of the present way of running economic activities is a potent force, which vested interests have to be reckoned with.

Among the promising developments in other religions I would like to mention the important work being done by Sulak Sivaraksa, a well-known social activist in Thailand and founder of the "International Network of Engaged Buddhists". (INEB) Sulak, sees religion at the heart of social change. In stressing the importance of spiritual values he is challenging an economic system, which grows through the cultivation of endless desire for consumption. This prominent Buddhist intellectual maintains that the real nature of a civilization can never be found in the cultivation of the desire to amass but only in the purification of human nature. Craving should be replaced by a longing for inner peace. A similar line of thinking is developing in Japan, notably in Rissho Kosei-Kai. This important Buddhist movement sees happiness determined by how well people have fulfilled their inner desires, by how close they have come to spiritual peace. "Even if consumption is one of the processes needed to bring people happiness, it is not the goal. The ideal is to be able to attain the greatest amount of happiness from the smallest amount of consumption."

The essence of this point of view was well summed up by E.F. Schumacher, author of the international bestseller *Small is Beautiful: Economics as if people mattered*. In this visionary book he states: "For the modern economist this is difficult to understand. He is used to measuring the 'standard of living' by the amount of annual consumption, assuming all the time that a man who consumes more is 'better off' that a man who consumes less. A Buddhist economist would consider the approach excessively irrational: since consumption is merely a means to human well-being, the aim should be to obtain the maximum well-being with the minimum of consumption."

Also interesting is the growing awareness among prominent personalities in the international world of finance and economics that there has to be a fundamental rethinking of the relation between religion and economics. A striking example is given by a former managing director of the IMF, Dr. H.J. Witteveen, in his writings on Sufism and economics.

It seems to me that the ongoing debate within Christian churches would be greatly enriched by a careful listening to those voices in Buddhism and Sufism, which are putting the emphasis on the achievement of "inner peace". One of these fruits could be the adoption of a more relaxed attitude towards consumption.

Economics—goal in itself or instrument to serve mankind?

From the foregoing it appears that religions could indeed provide valuable guidance for a new approach to economics. One of the first contributions could be a critical examination of the present way of handling economics. Are "economics" a goal in itself—as many think—or an instrument to serve the common good of mankind? For religious people as well as for adherents to other convictions of life it should be clear that the economy is only part of a larger Whole, not an autonomous entity ruled by absolute laws.

To avoid any misunderstanding I would like to repeat that my criticism of the present way of handling economics does not want to belittle its spectacular achievements. There can be no doubt about the impressive contribution of economics to the phenomenal improvement in the living conditions during past centuries.

The pressing question is, however, whether we have not been overshooting the mark. Surely, no one can object to sound economic acting and the optimal use of scarce resources for meeting material needs. But what is meant by "sound economics"? Are economics not degraded to a set of rules for maximizing profits instead of being a science serving the well-being of humanity? The iron laws of competition, efficiency and profit nowadays dominate practically all sectors. Profit, instead of service to humanity, is the driving force behind Science and Technology. Furthermore, who could deny that the much-praised invisible hand has led to misery for millions of poor and marginalized people?

The question about the purpose of economics is therefore of some relevance. A colourless definition says that: economics are dealing with production, distribution and consumption of material goods and services, as well as the art and manner of managing these different processes. These are words that are offering no perspective whatsoever in today's world. Hence, the need to develop a clear view on the essence of economics.

A few courageous economists have done pioneering work over many years in charting a new course in economics, often against all criticisms of their mainstream colleagues. Based upon their work I would like to submit the following definition:

The purpose of economics is the responsible use of the limited means at humankind's disposal in order to promote the common and individual well-being of present and future generations. Production, distribution and consumption of goods must be oriented towards a just and sustainable society in which the limits of nature are strictly respected.

Indeed, the essence of economics is to serve individuals and society in such a way that all people can lead a decent existence, within the limits imposed by the natural environment. Economics however are at present operating the other way round: mankind and society are serving economics! The original concept of economics as an instrument for the common good, for human's well-being, is perverted into a goal in itself, facilitating the enrichment of a limited group of people.

The above definition holds consequences, which will not be easily accepted in a post-enlightenment society in the grip of materialism and hedonism. Certainly not in a spiritual climate in which basic values like solidarity, compassion and respect for life are undermined. Economic theory and practice are nowadays built on a very narrow basis leaving little room for the basic teachings of religion. It is precisely this loss of vision on the place of human beings in the Ultimate Reality, which is at the root of the present crisis.

The forces resisting any change in the present situation are formidable. This makes it difficult to realize the profound transformation in production, distribution and consumption patterns now required. Not only commercial interests but also a large number of citizens show little inclination to accept a more sober life style. Material sacrifices are not popular in a secularized society where priority is given to the eager acquisition of material goods and immediate satisfaction of all sorts of desires. These reservations however will not withhold us from raising some fundamental questions about the driving forces behind the present economic order.

A critical evaluation of the myths behind the economic mechanism

In order to clear the road towards a just and sustainable economic development three persistent myths, now dominating modern society, have to be exposed:

- Human beings have unlimited material needs which must be satisfied
- Endless growth is a necessity
- We need a market without any constraints

Unlimited material needs?

The idea that human beings have unlimited material needs which have to be satisfied reflects an immature and irresponsible attitude. Immature because it is the attitude of a spoiled child. It is simply not true that human beings have unlimited material needs which have to be met in order to lead a satisfactory life. Adopting a mature attitude towards material goods is a matter of common sense. Welfare should never be confused with well-being. Many people in our affluent society have already gone through the experience that an increase in wealth—after a certain level is reached—does not lead to greater happiness. On the other hand we see that even a modest increase in income for the huge number of poor makes a great difference. In many cases the difference between . . . life and death.

Religious leaders and philosophers have emphasized throughout the ages that material goods will never meet our quest for happiness. They recommended instead practicing restraint, moderation and above all respect for life in all its forms. Buddhism even speaks about craving as the main cause for unhappiness. Indeed, humans infinitely surpass the voracious animal of a consumer society.

For Christians it should be obvious, that the insatiable appetite for goods in our present culture stands in shrill contrast to the Ten Commandments and the preaching of Jesus. The Gospel liberates us from the suffocating embrace of materialistic bonds thus redeeming us from the continuous pressure of social imitation of the fellow citizen. The teachings of Jesus Christ open a perspective towards a more realistic and promising approach. “Man is not to live on bread alone, but on every word that issues from the mouth of God." And in the Gospel of John we read: "Jesus said, I am the bread of life; he who comes to me will never be hungry, and he who believes in me will never again be thirsty . . ." In a society, so bent upon acquiring "perishable food", it would make sense to ponder upon the basic wisdom: "Work for no perishable food, but for that lasting food which means eternal life; the Son of man will give you that, for the Father, God, has certified him."

The concept of "unlimited needs" is also irresponsible in a world in which more than 1,300 million people suffer from hunger and misery. All those who are living in relative prosperity should realize that there is a direct relationship between a further increase in wealth in the economically highly developed nations and the poverty in a great part of the world. How can someone who is indulging in an oversupply of goods maintain in good faith to have unlimited material needs? Are we aware that our production and consumption patterns are creating serious negative social and ecological problems particularly for the poor countries?

Already the world is confronted with major ecological problems, largely due to the extravagant lifestyle in the prosperous North. It should be clear that our planet couldn't bear the additional environmental consequences of a similar economic development of several billion people. More than half of mankind is in urgent need of economic development in order to meet basic needs. Many are eager to realize a reasonable level of human existence. Preaching to developing countries that they should desist from doing so because of our concerns for the health of the planet will be in vain. Apart from being politically unrealistic it is also highly hypocritical in view of the extremely negative impact of our own environmental (mis)behavior. The concept of "unlimited needs" is therefore clearly an act of irresponsibility, degrading men and women to monsters of greed and lust. It should therefore be replaced by an attitude of enough is enough.

The present cult of consumerism is also doing a gross injustice to future generations. We will leave our children and grandchildren not only a degraded environment but also an impoverished world. They will be saddled with debts, which will most likely have to be met in a less favourable situation than at present. Debts, largely incurred because of the unwillingness of those living now to refrain from satisfying their "wants".

If we really believe in the basic solidarity between all inhabitants of our planet we—in the North—should reconsider our extravagant consumption patterns so that others—the majority of the world population—may live. The primary goal of economics should be to meet first of all the basic needs of all people. At the same time great care should be taken that wants are met within the limits imposed by social- and environmental conditions.

Endless growth?

Growth is inherent in nature. But not endless physical growth! Trees do not reach into heaven. At a certain point they stop growing in height; still they continue to develop. In a comparable way are human beings subject to limitations in physical growth while the process of inner development is supposed to continue. A healthy society should strive to achieve certain equilibrium between the inner drive to conquer new frontiers and the natural limits imposed by our planet. This basic wisdom is now foolishly pushed aside. In the dynamism of our actual economic growth model everything seems under an obligation to grow. This lack of common sense is amazing because permanent growth within a limited space is impossible. Even a child experimenting with a balloon could enlighten adults that one can't have unlimited expansion within a finite space.

Many people take for granted that "growth is good". But not all growth brings an increase in welfare. It all depends on what sort of growth is meant. New investments do not always lead to greater employment. Quite often these could even lead to a dismissal of workers. Jobless growth, caused by technological developments and profit hunting, is a real problem. The serious structural unemployment problem in our societies demands a new approach in which greater attention will be given to existing possibilities for employment opportunities in sectors relevant to the qualitative aspects in life such as health, education, social care, infrastructure and maintenance.

The urge for material growth in a finite world is nothing less than the growth of the cancer cell. Growth, in the economically highly developed countries, should be directed towards qualitative instead of quantitative goals. The present growth concept ignores the basic laws inherent to life on earth in order to realize a short-term material benefit. This is a fundamental error, as the laws of nature have to be respected if we want to prevent the undermining of the vital conditions for life. The need for growth in developing countries is indisputable; it is even an essential condition for the fight against poverty. But an additional claim from the highly developed nations on scarce natural resources is a sure recipe for disaster.

The underlying flaw in the present way of acting is that humankind today seems to have forgotten the basics of the human condition. Not only the Bible but also other sacred scriptures constantly remind us not to aspire to become gods. Human hubris, so typical of the growth concept, is indeed leading us into utter confusion and chaos. Hence the particular relevance of the story of the tower of Babel.

A completely free market?

The market is without any doubt a useful instrument for the allocation of resources and distribution of goods. The present emulation of the unrestricted free market however takes no notice of some serious limitations:

- Limited access. Participation is only possible for those who have avalable purchasing power. All others are excluded. This should be a matter of grave concern in a society with a great number of poor and marginalized people.
- The assumption of equal access of parties to the market seldom applies. A free market with a relative equality between the market forces is an exception. Market conditions are in practice controlled by powerful agglomerates and multinationals.
- The market does not register essential goods such as the quality of human relations, nature, beauty and silence. Goods, which have no price but make life worthwhile!
- The market does not function on the basis of a long-term vision taking into account the social and environmental interests of future generations. On the contrary, the market tends to operate on the basis of a short-term vision. A case in point is the low price of gasoline in the USA, notwithstanding the certainty that the present rate of consumption cannot be sustained over a longer period.

In spite of all these shortcomings the privatization of public utilities is being actively pushed through. For governments there are obvious advantages such as less current expenditures, elimination of risks and the worry of huge investment outlays. But the general public however is paying a high price. The original goal of public service at reasonable costs is replaced by the profit motive, which often results in less public service at a higher price! Private investors on the other hand will be able to benefit from the strong market position of the privatized enterprises. The growing tendency to privatization and liberation of the market system from all sorts of social fetters brings us every day nearer to a situation in which the law of the jungle prevails. Humans and nature are here subordinated to the iron laws of the market economy. The inevitable victims are among the increasing number of the poor and weak. A development, which runs directly against the ecumenical tradition of a radical choice for the poor and marginalized.

Religions should therefore be alert to the shortcomings of the market and come into action whenever human rights are violated or the integrity of creation endangered. They should be resolute on the unequivocal right on a human, dignified life for all people, irrespective of race or creed. Basic essentials such as food, clothing, housing, education, social and health care, should be assured for all. Not only for present but also for future generations. Religious teachings remind us that it is not we who should stand in the service of the economy but the other way round. Indeed, economics should serve the legitimate needs of mankind in a sustainable way.

The market should therefore always be embedded within a social and economic context.

Preliminary conditions

Although it is the main objective of this essay to focus attention on the contribution of the spiritual factor to a new approach in economics, some observations of a more practical nature have to be made. These are related to the following preliminary conditions which must be met if we want to set a course towards the goal of a just and sustainable economy:

- A reliable instrument for measurement of growth
- A more realistic cost and price structure
- Technological innovation

The first two conditions are obvious. We must know what the real impact is of our economic acting on human well-being and the natural environment. The third one is essential in order to stave off an imminent ecological catastrophe.

Creating a reliable instrument for the measurement of growth

The statement: "Economics aim at maximizing result," sounds fine. But there is one serious shortcoming—the GNP—the present instrument of measurement, is giving us a distorted picture of reality. Consequently we do not know what is going on and risk to count ourselves rich at the expense of real well-being. All sorts of elements—affecting our well-being in a negative way—are now incorporated as positive contributions to the Gross National Product. If car accidents increase and insurance companies pay higher sums for incurred damages, GNP goes up! Similar absurdities occur whenever increasing pollution or dense traffic leads to a rise in the sale of air and water-filters and the construction of expensive sound screens along highways. Ecological costs, connected with the production of goods, should not be ignored and passed on to others. These real costs should instead be deducted from the GNP.

Another example—of a different nature—is military expenditures—now absorbing a relatively large part of the national budget. Certainly, there will always be a need for an effective military force for peace operations. But do we really think that collective well-being will improve when the present excessive military outlays will be further augmented? Furthermore, how are the considerable environmental damages caused by military activities in peacetime accounted for in the present GNP? And what about the attribution of the extraordinary damages during wars of aggression, facilitated by the building up of a preponderate military position at colossal costs?

Politicians and economists continue—notwithstanding all these anomalies—to attach complete confidence in the accuracy of the present yardstick. Governments eagerly publish every year detailed accounts of the growth of the national product. These GNP figures look impressively precise to such an extent that even governments sometimes fall whenever growth remains a small percentage point below original expectations! Few people dare to question the sacrosanct nature of the figures. Truly surprising in these days of no-nonsense thinking and the adulation of a calculating mentality! If environmental costs would be deducted from the present GNP, there could very well be a negative growth in quite a number of countries now flattering themselves with attractive growth figures.

Clearly, without the development of a reliable indicator for the measurement of GNP it will be very difficult to set course towards a sustainable economy. We have to know what is really going on. In this respect attention should be given to efforts to compose a more reliable instrument for measuring the state of health of the economy. The Human Development Index of the United Nations Development Program (UNDP) represents an interesting effort to come to a more realistic approach towards a true assessment of welfare. Another intriguing example is the Gross National Happiness Index in the Kingdom of Bhutan. An original initiative, which merits serious reflection.

Towards a more realistic cost and price structure

Not only the present way of calculating the GNP but also the actual cost and price structure is presenting a distorted view of reality. Only cost factors expressed in monetary terms enter into the calculation. Ecological and social costs are usually externalized for the unwise reason that nature, air, water, beauty and silence are taken for granted and considered to be free goods. Obviously it would be more rational and responsible to insist upon the strict application of the polluter pays principle.

Similar mistakes are made with regard to "human costs". Dismissal of workers may lead to cost reduction and most likely to a higher profitability. But what about the "human costs" for the unemployed worker who is now being marginalized? What is the impact upon those around him or her? The presented cost figures therefore give a distorted picture of the real situation. A more sensible pricing of goods and services would be to internalize the full costs instead of passing these on to the general public—and . . . to future generations!

A distortion occurs also in the presentation of profits. Figures on the balance sheet may look highly promising. In reality however they reflect only part of the picture as the greater part of the "real costs"—ecological and social—are externalized. This means that while producers may register higher profits the public will have to foot the bill. Under these circumstances nobody should be surprised that human relations suffer. The negative social and psychological effects of acting on economic principles alone are regrettably still considered to be of secondary importance.

Technological innovation

In these days much emphasis is put on technological innovation, particularly in connection with the environmental crisis. Surely, much can be achieved by reducing the use of natural resources and the input of energy. Greater durability of products and recycling could also reduce the problem of waste disposal and pollution. The Wuppertal Institute, under Dr. Ernst Ulrich von Weizsäcker is working in this direction.

His book *Factor Four: Doubling Wealth, Halving Resource Use*[2], shows how twice the prosperity could be achieved with half the use of nature. Although I am convinced of the importance—yes, even the necessity—of technological innovation I seriously doubt whether this will be enough. The formidable environmental problems now confronting humanity also require—as has been repeatedly stressed—a fundamental change of attitude among the "haves". Without a greater sharing and caring on a worldwide scale, the present crisis will not be solved. Unfortunately, actual consumer and producer behaviour is pointing in another direction. The explosive increase in cars that consume high levels of fuel, is a case in point. It indicates that there is little disposition to take the energy crisis seriously.

We are therefore still a long way off before people will grasp that a profound change in mentality is required. Technological innovation is certainly needed but without a spiritual renewal our problems will not be solved.

How could religions support a just and sustainable development?

Religions are entrusted with a calling to orient hearts and minds towards the origin of our existence. They share a common responsibility to prevent a social and ecological calamity. But they will only be in a position to challenge the dominance of economics if they succeed in creating a much more intense and structured cooperation among themselves. In the actual world, dominated by powerful economic and financial interests, it is illusory to think that religions—operating separately—will be able to exercise a constructive influence. Without a bundling of resources they stand little chance to contribute effectively towards a just and sustainable economy. The strong resistance within secular society against any attempt to increase the influence of religious actors should not be underestimated! There has also been an inordinate increase of corporate and financial power in past decades. A power, which is practically unchecked. This absence of countervailing power is putting vital human interests of vast numbers of people at the mercy of a limited group of managers. This is—in view of the dynamics of the globalization process—creating a dangerous situation, challenging religions to formulate a common response in global thinking and acting.

Some practical steps

Acknowledging the responsibility of religions for a just and sustainable economic development is one thing; effective action is quite another story. High-minded statements from separate religious bodies on pressing economic and social issues will be of little or no avail in a world with strong vested interests. Viewpoints need to be presented in an appropriate way otherwise they will be met with a deafening silence by decision makers. Religions should therefore embark upon the venture of a much closer cooperation, enabling them to participate effectively in a joint dialogue with decision makers on economic, social and environmental issues.

A relevant step in this direction could be the formation of an Interreligious Platform on Economic and Social Issues in order to facilitate:

- Joint consultations on economic and social developments affecting humanity.
- Preparation of common positions on vital economic and social issues.

These positions should be based on a common analysis. A process, which could be greatly facilitated by a pooling of resources.

The Platform may decide to give special attention to:

- Environmental deterioration. The large-scale process of destruction and pollution of the natural environment now taking place as a consequence of a perverted concept of economics has to be stopped. The ruthless exploitation of human beings and nature—in striking contrast with the essence of religions—should be clearly exposed. The vision should be kept alive of a society in which love, justice, truth and beauty are not crushed under economic pressures.
- The worsening social situation both within the EU and between North and South. The present situation in which one-fifth of mankind is in dire need stands in shrill contrast to the persistent call of religions for justice and solidarity. Religions should accept full responsibility for effective action against poverty, unemployment and exclusion. Special attention should be given to relations with the economically less developed countries. This implies the revision of policies whenever these are having a negative impact on the weaker nations; particularly those dealing with agriculture, trade, development or finance. Attention should also be given to inconsistency in policies and economic practices leading to the crude exploitation of human beings.
- The implementation of recommendations and commitments of International Conferences such as the UN Conferences in Rio de Janeiro on Environment and Development, and the Social Summit in Copenhagen. The interdependence of economic and social development together with environmental protection was clearly recognized during these Conferences. Governments however have been rather reluctant to implement the recommendations and commitments notwithstanding convincing evidence about the serious state of affairs. It is therefore up to religions, as well as non-governmental organizations, to mobilize public opinion for further action.

In conclusion

Our present economic and social development is neither just nor sustainable. It stands in marked contrast to basic human values of justice, solidarity, peace and respect for life. Social and ecological disaster looms if there is no timely change in the irresponsibility of human behaviour. Technological developments, innovation, economic measures and fiscal policy are certainly of vital importance for a new approach. But unless these instruments are backed up by a strong motivating force—shared by wide circles of the population—these will not bring the desired results. The required change in attitude towards human beings, material goods and nature will not occur without a rediscovery of the religious dimension in life, a spiritual reawakening. Religious teachings hold precious—still undervalued—elements for a new approach to economics.

Religions are therefore confronted with a special challenge. They will fail to meet this challenge if they operate on an individual basis. None of them, individually, is in a position to challenge the actual dominance of economics. There is an urgent need for creating effective forms of cooperation! This may not only serve to prevent a further dehumanization of society but also create the conditions for a more just and sustainable development.

References

1. Donella H. Meadows, Jorgen Randers and Dennis L. Meadows, *Limits to Growth: The 30-Year Update* (White River Junction, VT: Chelsea Green, 2004).
2. Ernest Ulrich von Weiszäcker, *Factor Four: Doubling Wealth, Halving Resource Use* (London: Earthscan Pubs., 1998).

Hendrik Opdebeeck

Hendrik Opdebeeck *is a professor of Philosophy and Economics at the University of Antwerp where he is affiliated to the Centre of Ethics. He studied philosophy and economics at the universities of Leuven and Ghent where he received a PhD with a dissertation on E.F. Schumacher and the polyparadigmatic discourse in economics. His research interest is focused on the cultural-philosophical backgrounds and effects of globalization with special attention to the role of spirituality and technology. In 2000, along with Luc Bouckaert, he founded the SPES forum for Spirituality in Economics and Society. His publications in English include* The Foundation and Application of Moral Philosophy *and* Building Towers: Perspectives on Globalisation.

GLOBALISATION BETWEEN MARKET AND DEMOCRACY

Is the partition between efficiency and ethics justifiable?

Hendrik Opdebeeck

In this contribution, it is my intention to explore how one can prevent the dark sides of the globalisation process from leading to ever-greater ethical decay. By ethical decay, I mean specifically the threat that globalisation poses to what is referred to in the personalistic tradition as "the longing for being and solidarity with the other(s)". In this brief reconstruction of the core of the ongoing globalisation process, I shall take as a starting-point the text by Riccardo Petrella in the prestigious series of Grandes conférences at the Musée de la Civilisation in Québec, Canada.[1] The prevailing globalisation discourse manifests itself worldwide in political, socio-economic and scientific circles. It assumes both an explicative and a legitimising position vis-à-vis the globalisation process and its purpose. If one wants to put a stop to the decay of the longing for being and solidarity with the other(s), the urgency of a broader ethical (i.e. a more just) dimension in globalisation is more apparent than ever before. Of central importance here are the principles of merit and need as possible articulations of justice. Given the intrinsic logic of the merit principle, which induces a series of impermissible side-effects of globalisation, the traditional co-operation between democracy and market economy appears no longer able to provide guarantees for a just globalisation process. It is only through the integration of efficiency and a striving for more equality that one acquires insight into the need for appropriate global institutions.

1. The Discourse of Globalisation

The current globalisation of markets, enterprises and capital goes hand in hand with an ongoing revolution in information and communication technology. We need to extrapolate to where these technological developments will lead if they are allowed to pursue an autonomous course. It is claimed in the globalisation discourse, which manifests itself in all languages and via all possible media, that globalisation is an inevitability against which nothing or no one can put up meaningful resistance. Thus, the key notion is adaptation: one appears to adjust at all levels to this process. Anyone who fails to adapt is excluded, if only on the basis of the hard rules of competition. The imperative of global competition of all against all is paramount. The new technologies are put forward as the most powerful and efficient tools for guaranteeing competition on world markets. The omnipresence of this discourse stems from the *de facto* economic power of the most developed nations, whereby the values and the criteria that are characteristic of the capitalist market economy are elevated to exclusive global values and criteria with regard to what is good, useful and necessary. This is apparent not merely in the economic domain, but in any possible field. Hence, the purpose of the history of contemporary society is equated to a necessary, inexorable and inevitable evolution towards a single and self-regulating global market place. The nations of the world have set themselves the principal task of integrating their national economies as smoothly as possible into the global economy, or at least of not imposing obstacles to this integration. Thus, they have relinquished their own political accountability. Everything is subordinated to this finality: investment policy, technological innovation, labour market policy, education, trade regulations, tax policy and—in the case of Western Europe—the process of European unification. Scientific and technology policy, including the stimulation of R&D, is one of the most striking examples of this trend. Consequently, a new alliance is formed between government and private enterprises on the free market, for the purpose of realising their common goal of a competitive integration of the national economy into the global economy.

The discourse of the inevitability of an integrated global market also explains how it has been possible over the past twenty years for the economic and socio-political system to experience a thorough reorientation on the basis of three principles: liberalisation of markets, deregulation of the economy, and privatisation (not only of banks and railways, but also of hospitals and schools). It is quite striking in this respect how political deregulation has made way for financial regulation. Today, finance is no longer controlled by politics, but by the money markets. This evolution has unfolded against the backdrop of an ever-greater divergence of the financial-economic and the real economic spheres. Only a small fraction of current financial transactions are aimed at the creation of real new wealth. The rest is purely speculative.

A final important characteristic of today's globalisation process as described by, among others, Petrella is the imposition by the greatest economic powers of social, democratic and environmental clauses in the new regulation of the World Trade Organisation (WTO). These clauses concern the quality of labour, human rights and environmental standards respectively. While these are undoubtedly important clauses, the question remains whether the WTO is the most appropriate body for controlling global social, democratic and environmental standards.

2. Globalisation and the Decay of Ethics

In order to ascertain whether or not the globalisation discourse described above entails a decay of ethical striving, it is worth recalling what, according to Paul Ricoeur, is the foundation of "the longing for being".[2] The starting-point for Ricoeur's theory of ethics is his assertion that the human being constitutes an original confirmation of being, which manifests itself in endeavour and longing.

What Ricoeur means by original confirmation of being is the act whereby we are placed in existence, which precedes our thinking, acting and feeling, and underpins and guides them, but which is never given to us and can therefore never be beheld. One could, therefore, regard this original confirmation of being as an act of freedom; the basis of what we are and what there is for us to be. So Ricoeur wishes to lay bare the human being's longing for being, in order to help him or her acquire a more adequate degree of self-understanding. This is the only way that a more authentic (co)existence can be attained that transcends alienation.[3] Ethics thus enables us to identify the meaningful history of our longing for being. Clearly, then, the notion of freedom is central to Ricoeur's theory of ethics.

Certainly in the light of the constantly developing free market economy within the globalisation process—as a kind of structural anchoring of the central ethical concept of freedom—Ricoeur's perspective is extremely interesting. Note, however, that the notion of freedom is not so easily grasped. Therefore, reflection is required on its concrete expressions or manifestations. The above hermeneutics of the globalisation process ties in with this interpretative process.

Significantly, the act of freedom can be considered as prior to evil (as contested by the anti-globalist movement). According to Ricoeur, it is however the case that evil and freedom elucidate one another and, indeed, that evil not only mutilates ethical striving at the individual level, but also within the structures by which we operate. Ricoeur formulates this very clearly when he asserts that there is a specific evil of the institution, i.e. the evil of objectification that one encounters in all forms of organisation. In the distribution of labour, for example, it assumes the subtle shape of drudgery, which slowly takes possession of hyper-specialised and monotonous work in industry. The erosion of the right to meaningful employment in the globalisation process ties in with this view entirely. Those in charge of our economy are calling louder and louder that it is no longer possible to provide full employment, let alone meaningful and qualitative employment. In the European Union alone, there are close to 20 million unemployed. Worldwide, the number is estimated at around one billion. Fundamentally, what is at issue is that people are no longer regarded as individuals, but as a human resource, comparable to natural, technological and financial resources. If the cost of labour becomes too great, workers are made redundant and they are replaced with another economic resource, such as a machine. The evil within the globalisation process is therefore an unfathomable manifestation of freedom that makes freedom itself unattainable, and that immediately restricts our ethical striving.

3. Globalisation as a Threat to Solidarity

If the starting point of ethics is that of the free human being, longing and striving for being, then one only attains actual ethical existence through the encounter with the other.[4] This recognition of the other as a person at once entails an obligation. The freedom with which everything started, including the free market of the globalisation process, is also recognised in the other person.

Of fundamental importance here is that Ricoeur extends this intersubjectivity to the whole world through the process of norms and laws. He renounces individualist ethics, because a person is not merely an individual, but constitutes various levels of the 'we', which are embedded within the totality of structures and institutions. In this respect, Ricoeur's ethics is reminiscent of the ideas of such personalists as J. Maritain and E. Mounier,[5] to whom the person, as an individual and a collective entity, is the norm.

Considered in this context, Petrella's warning that globalisation is increasingly threatening solidarity between the 'I' and the 'other(s)' is of great importance. He asserts that the phenomenon of growing poverty is not only an indication of the negation of the solidarity principle in the welfare state, but also of the negation of citizenship. Indeed, it is increasingly argued that citizenship does not imply social rights of the other. In the world's richest country, the United States, 1 in 5 people are living in poverty. In the European Union, the proportion is about 1 in 6, but in the UK it is as high as 1 in 4. This also manifests itself in an unequal income distribution. In the US, one percent of the population possesses 40% of the national wealth. This unequal distribution of wealth is manifesting itself in the emergence of more and more private neighbourhoods in cities, where the rich can cut themselves off from the other(s), and enjoy extra protection against violence and crime. It is also striking that technology is one of the principal factors in the dynamics of social exclusion. With respect to work, for example, one sees how difficult it often is for the unemployed to re-enter the labour market, since the greatest possible number of jobs are being made redundant through technology.

Moreover, the present economy is increasingly making us lose our sense of being and of working and living together with the other. In the light of the global competitive struggle, solidarity has become an obstacle and an unbearable cost for enterprises. This is partly why the value of the common interest is being lost. Priority is given to individual careers ('my education'), to individual survival strategies ('my job and income') and to individual possessions ('my home, my car') as fundamental and irreplaceable expressions of freedom. Sight is lost of the other. The logic of economic competition is made into the obligatory logic of society as a whole, which inevitably results in a victory of the one over the other, or, if you will, in the exclusion of the one by the other. Particularly the less developed countries are worse off for the process of globalisation. A mere 15% of global direct investment is not intended for the most developed economies.

4. Globalisation and the Need for Adequate Global Institutions

When a free individual enters into an ethical existence through the encounter with the other, he or she can only do so by mediation of institutions. A third aspect is therefore required, i.e. a neutral term, an institution. Hence, the threat that the globalisation process poses to the "longing for being" and "solidarity with the other(s)" increasingly necessitates an appropriate global institution. An institution consists of a set of rules for acts in social life, whereby each person can realise his or her freedom without impairing that of others. It is within the framework of existing institutions that human beings act. These institutions are not detached from value judgements—as in the discourse of globalisation—so that they are already qualified ethically and, as such, can either bring forth freedom or not.

In this context, Petrella points out that globalisation, with its drive towards liberalisation, deregulation and privatisation, has drastically reduced the significance of public power as an institutionalisation of freedom. This finds expression in, for example, the waning influence of parliament. Government is increasingly giving in to the demands imposed by the, often-aggressive, markets. The most detrimental aspect of this evolution is that, because of the growing tendency on the part of national economies to adapt to the global economy, the public authorities have failed to develop a global political authority; on the contrary, this international integration actually enhances private power at global level. This is also true at the European level. The more that member states apply the principles of liberalisation, deregulation and privatisation, the further they appear to remove themselves from developing a federal European power. The fundamental question here is what is meaningful freedom?

Ethical mediation must ultimately be beneficial to the human person if one is to avoid bureaucratic stalemate. Indeed, the very purpose of institutions is that they should serve people; if not, institutions are entirely meaningless. Certainly in the light of the globalisation process, the big question is whether specific ethical rules need to be established at the institutional level, since it is difficult to extend those from the interpersonal sphere to institutions. This explains the great significance of justice as an institutional tool that allows different freedoms to co-exist. Following in the footsteps of Aristotle, Ricoeur further elaborates this assertion by underlying the connection between justice and equality. However one interprets equality, its meaning for institutions is what care-for-the-other or solicitude means to interpersonal relationships.

5. Justice Criteria

Justice has always been a moral characteristic whereby people grant others, with an inner self-evidence, their rights.

Justice relates not only to property, commodities and finance, but to all values that people must realise with and for each other, including safety, health, marriage, education, truth, freedom and life itself. In fact, it concerns a broad range of issues, much like those raised by the anti-globalist movement.

The 18th-century philosopher David Hume refers to four circumstances which make justice indispensable in our society.[6] First, the individual is required to co-operate with others in order to satisfy his or her personal needs. Surviving as an isolated individual is impossible. Second, the extent of a person's benevolence is limited, so individuals are quickly inclined to claim everything for themselves, without taking account of the other. Third, goods are scarce, so that conventions are required to establish right of property. Finally, all human beings have more or less the same basic needs. Satisfying these needs presupposes a reasonable distribution. Otherwise one will be reduced to inhumanity, as the anti-globalist movement again argues forcefully.

This brings us to the confrontation between justice and excessive economic power, as in the present context of globalisation. Abuse of economic power at the level of communal justice manifests itself as exploitation, consumer manipulation and market monopolies among other things. But economic power also plays a role in legal justice. Examples that spring to mind are various forms of tax evasion and the bypassing of government by, among others, multinational corporations. The issue of globalisation also comes into play in the field of distributive justice, which is concerned with a reasonable or equitable distribution of scarce commodities.

There are two important principles of distributive justice in the economic domain, these are "grant everyone their share according to merit" and "grant everyone their share according to their needs". These two principles not only typify the age-old tension between liberalism and socialism, or between capitalism and communism, but, as will become apparent, they also shed light on the origin of injustice in the present globalisation process.

The notion of merit has different meanings. It can refer to a performance or a responsibility, as well as to a mental effort. The principle "grant everyone their share according to merit" always refers to the past, as merit is inevitably based on assets such as talent, skills or knowledge, all of which were acquired (unequally) in the past. Consequently, this criterion tends towards inequality.

An example that springs to mind in the context of globalisation is the important aspect of income distribution. This is almost a reflection of the inherently unequal distribution of, for example, intelligence among the population. As the logic of globalisation attributes such a central role to earnings, one obviously aims at increasing economic output quantitatively and qualitatively by stimulating performance with a view to realising the greatest possible utility.

As regards the second important principle, i.e. distributive justice by "granting everyone their share in accordance with their needs", this refers to the most essential needs of human beings. These needs are inherent in "being human", and they include the need for food, clothing, housing, etc. If one interprets justice in the sense of granting to the other whatever the other is entitled to, then this second principle may offer a guarantee for realising more social justice.

On condition, that is, that one considers intrinsic basic needs that are a genuine prerequisite for a dignified human existence.

6. The Failure of the Mixed Market Economy

Certainly in business circles where people are sensitive to criticism of globalisation, it is usually claimed that it is impossible to realise both principles of justice in the economy. The application of both the principle "grant everyone their share on account of merit" and that of "grant everyone their share on the basis of their needs" would result in a short circuit, if only because the merit principle creates inequality while the needs principle does the opposite. The anti-globalist movement tends to argue in favour of the needs criterion. This in itself is no problem according to proponents of globalisation, but the latter also claim that those same anti-globalists wrongfully accuse the free market of not applying the needs principle and overemphasising merit and pay according to work. They argue that the anti-globalists tend to lose sight of the fact that the market economy is based on the performance criterion rather than the needs criterion. Consequently, they denounce familiar economic power mechanisms: they reject excessive economic competition, propose rather drastic redistributive measures, criticise multinationals, disapprove of the stimulation of consumerism (through advertising), deplore the substitution of capital for labour, and suspect the market of wanting to reduce humans to the status of slaves of commodities. Therefore, in the eyes of the anti-globalists at least, it is justified to give the state more of a *carte blanche*.

The traditional proponents of globalisation argue that, if one does want to take into account both the merit and the needs principle, without imposing these criteria on either the economy or globalisation, then it is essential that the economy and politics should be made to function as efficiently as possible. For this reason, they defend free competition and true democracy, claiming that they serve the general interest. The merit issue is resolved on the basis of the free (i.e. economic) market mechanism, while the needs problem is approached from the perspective of democratic (i.e. political) decision-making. The general interest objective realised in this manner brings forth the greatest possible degree of justice.

According to the champions of globalisation, this at once shows that human self-interest, to which the profit principle is so often equated, is not entirely negative; on the contrary, self-interest may be the very foundation of the general interest. However, if this mechanism is to be sustainable and not lead to collective self-destruction, then moral characteristics, such as sympathy and solidarity, mutual trust and justice, are required. Self-interest and the general interest would then complement each other in much the same way as two buttresses carry the vaulting in a cathedral.

Globalisation and justice, self-interest and general interest, the merit principle and the needs principle, free competition and true democracy. According to the proponents of globalisation, each of these pairs can be combined most effectively in a mixed market economy, as this system allows one to reconcile the production issue (cf. efficiency) with the distributive issue, where satisfying basic needs is of primary importance. Indeed, a public authority is necessary in order to guarantee that the degree of private power within a national community remains within the general interest.

7. An Ethical Dimension in Globalisation

While a mixed market economy would appear to offer a logical solution to the tension between globalisation and justice, the question remains whether this partition between (economic) efficiency and (democratic) justice, or between globalisation and ethics, is justifiable in this day and age. After all, the merit principle that is central to the economy not only causes inequality, it also stimulates economic growth and progress, unlike the needs principle, which tends to restrict growth. However, in recent years serious side effects have been occurring in relation to economic progress. It is apparent from the emergence of the anti-globalist movement, that there is growing awareness of the end to growth and progress. It has become very clear that, if the industrialised Western countries want to avoid undermining themselves, the emphasis in economic growth needs to shift from quantitative to qualitative growth. One needs to abandon the traditional concept of growth, as it results not only in a disproportionate appropriation of scarce resources but also in irreversible damage to the environment. This conception is contrary to the policies pursued in many Western countries. All too often, economic growth is seen as the prime objective, to be achieved by a market that is unimpeded by government. Thus the most pressing problems facing the world today are transferred unscrupulously to the environment, to the third and fourth worlds, and to the quality of inter-human relationships. The question "What is the sense of it all?" resounds loudly in Seattle, Genova and Laken.

Indeed, in this manner, economic power enters the triangle of profit, progress and needs, rather than merely confronting the issue of justice. Goudzwaard[7] has long asserted that, in the iron law of so-called economic and technological progress, which forces everything and everyone to constantly adapt, we have discovered the secret we were searching for. In an economic order where progress occupies a central place, wielding of power can indeed develop in only one direction: that of ever-greater productivity increase and a matching standard of living. This compelling force can apparently afford to neglect the negative power effects of globalisation (Third World, unemployment, environmental degradation . . .).

In view of the negative effects of growth, as articulated in the dynamics of globalisation, striking a balance between democracy and market economy (i.e. the mixed market economy) is no longer a guarantee for realising a just globalisation. Due to the intrinsic logic of the merit principle, which not only induces inequality and economic growth, but also impermissible power effects, the boundaries of globalisation call into question whether this exclusive principle is sufficient in the economic field, let alone acceptable.

Consequently, more and more people are calling for a broader ethical dimension in globalisation (rather than ethical decay). This way, one can prevent the one-sided merit principle continuing to occupy a central position in the globalisation process. It is striking that these calls are now heard right up to the highest levels. Participatory justice is considered to be of the utmost importance. One goes beyond the two principles of distributive justice and abandons the (dualistic) opposition between the individual and the community in order to arrive at a shared responsibility and involvement at national and international level.

8. A Global Social Contract

The concept of a global social contract is thus an important elaboration of the concern with the increasing equality of institutions. Petrella quite rightly asserts that the essential challenge for the global economy lies not so much in the integration and adaptation of the national economies to the world economy. It lies in adequately determining the principles and rules within the institutions that are required to provide the world's population with its basic needs in terms of food, housing, energy, health, education, transport and employment. The central question is, in other words, how can one achieve global wealth rather than predominantly private wealth at a global level? Or how can one assume responsibility in true solidarity instead of developing one-sided competition?[8]

The main difficulty in realising a global social contract is, however, the virtual lack of a global political system that can shoulder the responsibility of protecting the general interest.

Clearly, politics needs to assure the survival of the community; all the more so if its future is under considerable threat. This acceptance of responsibility implies caring for the most brittle and the most threatened thinkable. With this we have arrived at the philosophy of E. Levinas: we are ultimately responsible for the other(s), without thoughts of reciprocity and without room for calculation. The purely reciprocal recognition between people is however transcended in order that everyone would assert himself in individuality, which is in fact the starting point for Ricoeur's ethical genesis. According to Levinas the egocentric longing for being of the self is only transcended when the 'I' is obliged to do so, and, more in particular, when it responds positively to the appeal of the Face of the Other.[9] When Ricoeur discusses the recognition of the other, there would appear to be a certain ambiguity. Indeed, one can restrict oneself without questioning oneself, without insight into the arbitrariness of the own (despotic) longing for being. The auto-limitation is founded upon the narrow base of the war of all against all, or the threat that the other poses to one's own freedom. The self does not (yet) perceive its own egocentrism as unworthy of a human being. Thus, it would appear that Ricoeur runs the risk of reaching a dead end at the level of well-understood reciprocal interest. This is probably due to his perception of the other as an *alter ego*: *alter* certainly, but still an *ego*. But meanwhile an ego-centric globalised society has been established, which in social, economic, financial and legal terms takes the shape of egocentrically-organised political structures at national, international and global level. It would appear that precisely these structures are called into question by the anti-globalist movement today.

Notes

1. R. Petrella, *Écueils de la mondialisation. Urgence d'un nouveau contrat social*, Montréal, Éditions Fides, 1997. Other treatises on globalisation in the same vein as Petrella include F. Buelens, *Globalisation and the Nation-State*, Cheltenham, Elgar, 1999; W. Derkse e.a. (eds), *In Quest of Humanity in a Globalising World*, Leende, Damon, 2000; J. Scholte, *Globalization. A Critical Introduction*, Basingstoke, Palgrave, 2000; N. Hertz, *The Silent Takeover. Global Capitalism and the Death of Democracy*, London, William Heinemann, 2001; N. Klein, *No Logo. Taking Aim at the Brand Bullies*, Toronto, Knopf Canada, 2000 and I. Ramonet, *Géopolitique du chaos*, Paris, Editions Galilée, 1997.
2. For a more extensive overview of Paul Ricoeur's ethics, see H. Opdebeeck (ed.), *The Problem of the Foundation of Ethics*, Leuven, Peeters, 2000.
3. Compare with E. Weil, *Philosophie politique*, Paris, Vrin, 1956.
4. P. Ricoeur, *Soi-même comme un autre*, Paris, Seuil, 1990, pp. 212-213.
5. E. Mounier, *Révolution personnaliste et communautaire*, Paris, Aubier, 1935.
6. D. Hume, *A Treatise of Human Nature*, Oxford, Clarendon Press, 1978.
7. B. Goudzwaard & H. de Lange, *Beyond Poverty and Affluence. Toward an Economy of Care*, WCC Publications, Geneva, 1995, p. 92.
8. It is clear that R. Petrella is on the same wavelength here as J. Rawls, *A Theory of Justice*, Cambridge, Mass., Belknap Press of Harvard University Press, 1971.
9. F.i. E. Levinas, *Autrement qu'être ou au-delà de l'essence*, La Haye, Martinus Nijhoff, 1974.

The Arts

David Keplinger

David Keplinger *is the author of four poetry books, most recently* The Prayers of Others *(2006) from New Issues Poetry and Prose. He has received grants and awards from the Soros Foundation, the Katey Lehman Foundation and the Pennsylvanian Council for the Arts. His first collection,* The Rose Inside, *won the 1999 T.S. Eliot Prize. Keplinger is associate professor of English at the American University in Washington, D.C.*

TWO VISITS: TWO LESSONS

DAVID KEPLINGER

Lessons From Rome: Keats and the Weight of the Real

Just off the Spagna Metro stop in Rome are the Spanish Steps, where for centuries students have lounged in the spray of Bernini's gigantic fountain, whose Neptune holds a sinking ship—momentarily, eternally—above the waves. In its shadow sits the Keats house, a thin and unassuming domicile with winding staircase ascending three floors. Here Keats and Joseph Severn lodged the last three or so months of the poet's life. Sheer luck had drawn me to it, turned my head in that direction, though somewhere in my memory I know I'd heard of its existence. Opened to the public only in the last four decades, the house is preserved to a replica of its early 19th-century charm. I charged at it as if I'd spotted an old friend across the piazza. Up the three floors and I was entering an antechamber leading to the room, facing the square, in which Keats died, the minuscule bed on which he had lain miserable and sick. I found myself looking down from the very window Keats had watched these self-same unchanging faces of the young, of which he was one. When I turned I saw, to my astonishment, the original death mask made from plaster on the night he died, February 23, 1821.

I had written a poem about the mask for my second book. I stood in wonder at this artifact which had hung here all along as I wandered twenty years ago, and this morning, through the square. I was astonished by the youthfulness of the face, mixed with a sinking defeat that I sensed in the sunken eyes, the hard lines of the mouth.

The museum provided maps to the grave site, the Protestant Cemetery very near the Pyramid stop on the Metro. I could see the grave in my head: those disturbing lines on his epitaph ("Here Lies One Whose Name Was Writ on Water"), dedicated anonymously, as he had requested, to "A Young English Poet." Unlike Shelley's epitaph, just above Keats' on a hill in that cemetery, and which promises his transformation into "something rich and strange," Keats' grave offered no such assurances. It offered nothing but the fact: One's life is short and mutable, dissolving. His death, though hard, also hardheadedly embodied the life of a man "capable of being in uncertainties, Mysteries, doubts without any irritable reaching after fact & reason."

In *Keats in Rome*[1], Nigel a Brassard recounts Keats' last days:

> On February 12 Severn acknowledged Keats' desire that "the hope of death seems his only comfort." A sentiment which echoes "Ode to a Nightingale" "for many a time, I have been much in love with easeful Death." Two days later Keats asked Severn to visit the Protestant Cemetery and requested that the unopened letters from Fanny, a purse made by his sister and a lock of hair given to him by Fanny, would be placed in his coffin. Keats asked that his gravestone should not have his name only the words "Here lies one whose name was writ on water." That night, staring at the floral pattern on the ceiling, Keats said he felt the flowers growing over him.[1]

Keats died in Severn's arms. "Don't breathe on me," he told Severn. "It comes like ice." The grave sits next to Severn's, who lived to 85. Like an old married couple, between them is the little stone remembering Severn's young son. At 85 Severn must have thought of Keats a father and a grandson all at once. And that epitaph without a name! The reference to water. I thought of the fountain by Bernini, in the center of Piazza di Spagna, which Keats saw every day, a boat sinking in the spraying waters.

How mysterious that Keats' whisper has only compelled us to bend in more closely, to press our ears up to his dying face. In his essay, "Beauty, and/or Truth," Philip Levine speaks of Keats' concept, *The Vale of Soul Making*. "Through the agency of his letter to his brother George on the world as the *Vale of Soul Making* I felt myself becoming a religious person," Levine writes. "He goes on to tell George and us that the world is the horn book from which each intelligence learns to read; that is, to become what it is capable of becoming, a singular identity, a soul; without the education our experience of *this* world can give us we remain less than a soul, merely a potential."[2] To gain a soul the cost is very great in Keats. You have to lose your worldly eyes, lose your illusions about life by entering life. That means recognizing that you are no more special than anyone else. Your name is writ on water, too. But there's a twist. Keats' grown-up insistence we forget his name has made him renowned. A poem—even the surrealistic—founded in such realism must accept life on life's terms. Robert Bly has called this the necessary quality of "adult grief" which ultimately strengthens the writer's authority. Poets reach their height in their sixties and seventies because that's when the fact bears down. You're really going to die. It happened early, much earlier, in Keats. But the outcome is the same.

Keats came to Rome a dying man. By early January he rallied, though, and even went for walks and, once, a ride into the city for some fresh air. But then the relapse came, and he was bedridden for the last month. He lay in the single bed at Piazza di Spagna 26, on Pincian Hill, the Tiber River out to the west. On the night he died they cast the death mask. I had found a photograph of it, and it became the object of obsession for weeks. I wrote hundreds of poems, but finally an 18-line piece using a loose form from Auden. The last stanza goes

> Already late in February, the famous
> Roman earth kept opening; the bees would grow colossal
> With alacrity and greed. Without him in the world,
> The world had failed to change. What in heaven would?
> His last report's a homely fossil
> Ever young; the eyes shut up; ever knitted to their darknesses.[3]

Others who came found the image of Keats, his grave, his life, a compelling commentary on *The Vale of Soul Making*. He was taken to the grave which now lies twenty yards from a busy intersection near Pyramid, where Oscar Wilde visited in 1882. Sitting on the stone bench, Wilde wrote,

> Was thy name writ on water? It shall stand
> And tears like mine shall keep thy memory green
> As Isabella did her basil tree.

For two centuries now, countless writers have come to the Protestant Cemetery where, as the traffic raced just ten or so feet beyond the fences, this grave, with its lyre relief and encrypted epitaph, stands. And that skeleton boy just under me, the last letters from his betrothed, Fanny

Brawne, curled up in its skeleton fist, letters which he never opened. (He claimed he couldn't bear to read them.) The grave and Keats' life and the friendship with Severn, and Severn's love for his friend, and the poetry, it was all art, fused beyond life and art and being both and neither. Keats at a very young age had found his way in. And all the Oscar Wildes and poets who had come here had been seeking a way in. The secret is: Forget your name. Forget fame. Forget publishing. Lack the "irritable grasping toward reason" in the forests of uncertainties. Live life on life's terms.

At the grave one of the tens of cats of the cemetery, a descendent of the very cats of Keats's Rome, fluttered in and out between my ankles. It was purring so loudly I could hear it before I bent down. It was a scrappy thing. At my touch it maddened with delight—the purr got louder; it nearly fell down from pressing so hard into my ankle—and in that "theater of responses" I felt the mounds of ticks beneath its fur and saw that it had a continually running nose, a pus-filled infection actually, which caused it to sneeze every time it breathed in hard my scent. This cat, infested, infected, was sick; but what a welcome it gave! As if merely being were enough for gratitude—and this reality, exactly as it is, the gift.

Writing as Alchemy: Lessons from the Land of St. Francis

1. The Mystical Component

We hear William James' definitions of mysticism and religious experience often because, for one, they provide a safe academic haven in which to talk about these mysterious events. Psychiatrists might call a mystical experience a psychotic episode: voices are heard, etc. But James did not. He treated them as life-changing phenomena. In many cases, the voices are particularized to a certain moment in time and do not come back, as they do for schizophrenics. This makes the message of the voice more poignant, for sure, but it's a mystery: where on earth, if not the dissociated psyche, did that voice come from? James lists the four criteria for such an experience, and I think it is worth our while to discuss them here. Each experience (and he cites many accounts in *Varieties of Religious Experience*; there is a chapter called "Mysticism") is marked first by its ineffability. There is no way to properly describe what has happened. When Dante emerges from Paradise, called to write down his journey, his first response is wordlessness. The ineffability of the mystical experience recalls its directness: such an experience is just that; it can't be taught.

James says, secondly, that each experience carries with it a noetic quality. There is some illumination that comes of it; some learning takes place. When Francis hears the voice (of Jesus) demand, "Repair my church," he first takes that literally and steals some expensive tapestries from his father's home and sets out to reconstruct San Damiano. But it is a metaphorical message, more. Still, it is a message, the transmission of knowledge from some outside source. The third characteristic is transiency. These experiences do not last very long. They cannot; for in them the connectivity that exists between all things, we are told, appears to the visionary in luminous clarity. It's like staring into the sun. And finally, James tells us that passivity seems a shared quality among these visionaries. We may manipulate our bodies to get there, with drugs or breathing exercises or starvation in some cases, but once inside the experience it would seem that a power greater than ourselves is in control. We think immediately of visionaries like Blake or Dante, who claimed such experiences as elemental to the writing process. The body becomes a vehicle for that higher power. For Blake, who, by all accounts, was a pretty regular guy (Eliot writes, "There was nothing of the superior person about him. This makes him terrifying."), these visionary experiences happened all the time. Thomas Stearns seems a bit jealous in his essay on Blake. He writes, "Blake's poetry has the unpleasantness of great poetry." And, in a famous passage from Eliot:

> What his genius required, and what it sadly lacked, was a framework of accepted and traditional ideas which would have prevented him from indulging in a philosophy of his own, and concentrated his attention upon the problems of the poet . . . [and this is] one of the reasons why Dante is a classic, and Blake only a poet of genius.[4]

To Eliot, Blake's lack of "accepted" frameworks makes him a failure. Blake's unique philosophy is not an indulgence, I would argue. It is the result of James' third criterion, that noetic quality of mystical vision. The framework emerges from a higher power (call it Imagination or God), which, though unique to Blake, was not personal. All religions began, Christ's included, as heretical philosophies. That Blake's philosophy is original and heretical makes it more believable to me. So too was the case with Francis of Assisi. His two famous texts, *The Book of Life*, a series of prayers, and *Canticle of the Creatures*, remain staple literature within the Catholic library today. You may have remembered a few lines from Francis without knowing you did. "Make me an instrument of your peace," he pleads to God. "Where there is darkness let me be the light . . . For it is in forgiving that we are forgiven, and in dying that we receive eternal life." Francis' aim was to completely dissolve into God. He wanted no credit for any of his acts, but only to be the channel for right action. By the time he had reached his twenties, he abandoned all, to his parents' dismay and horror, for the life of a pauper, donned a brown sackcloth and tied to his waist a piece of rope. His father had been a rich textile merchant! But this is the image we have associated with him, and his plea for asceticism, for 800 years.

Francis is the patron saint of animals. A poet, he wrote *Canticle of the Creatures* at the Santuario di San Damiano, where he first heard the voice of Jesus, about a mile south of the city.

Praise be to You my Lord
For Sister Moon and for the stars
In Heaven You have formed them,
Shining and precious and fair.
Praise be to You my Lord
For Brother Wind
For air and clouds, clear sky
And every sort of weather . . .
For Sister Water
She is useful
And humble, precious and pure.

This anthropomorphism of the wind and water and the moon are sooner to be found in Native American drumming ceremonies than they are in a tradition like Catholicism, which passes over the natural, fallen world for the divine. Francis' confidence and knowledge came out of his deep vision. And, while his poem is pretty darn good on its own, it carries the authority of one who has experienced the world in this familial way. He is not writing, as Eliot would suggest he do, from a framework of previously accepted beliefs. It's the uniqueness of his vision that characterizes its authenticity.

Though Eliot has heaped Dante into the "classical" section, there is something anti-classical about Dante's visionary system, too. For Dante, vision's noetic quality had to do with love; a physical love that deepened into something spiritual. First he sees the other; then he sees that the love he experiences is the true reality of his existence, and, folding back towards himself, he falls in love with his own life. In his great *La Vita Nuova*, a book to which I owe a large portion of my education as a poet, he recounts his childhood vision of Beatrice:

> As I thought of her I fell asleep and a marvelous vision appeared to me. In my room I seemed to see a cloud the color of fire, and in the cloud a lordly figure, frightening to behold, yet in himself, it seemed to me, he was filled with marvelous joy. He said many things, of which I understood only a few; among them were these words: *Ego dominus tuus* [I am your master]. In his arms I seemed to see a naked figure, sleeping, wrapped lightly in a crimson cloth. Gazing intently I saw it was she who had bestowed her greeting on me earlier that day. In one hand the standing figure held a fiery object, and he seemed to say, *Vide cor tuum* [Behold your heart.][5]

The Master of the vision is the alchemist, so to speak. He holds this fiery object (the heart) that will bind Dante to his Other. The heart, in alchemical terms, is the philosopher's stone, required for the transformation of lead into gold. Love is the fire that surrounds the heart. Only love of this reality, on its terms, makes it possible to put ourselves in the place of another person. Without it, we're alone with the fact: the brain is enclosed in this body. When we lift the veil of self-protection, love is possible. After the vision, the visionary is changed for life. Dante was changed. Francis was changed. They had a confidence they lacked before. We all know how it feels to be on the fringes of love. Am I really in love with this person? we ask ourselves over and over. And we know what it is like to have an experience that pushes us into a corner—a near-death experience, say—and we come away from it with the sense that the world is more clear, our priorities are more clear. We either end or stay with the relationship, and we do either one with authority. That's the power of vision. It moves one forward. It's what Hamlet lacks (though Hamlet is having visions of his father all the time). "I lack advancement," he tells Rosenkrantz and Guildenstern. In James' words,

> The keynote of it is invariably a reconciliation. It is as if the opposites of the world, whose contradictoriness and conflict make all our difficulties and troubles, were melted into a unity.[6]

2. Melting With the Other

This melting James describes, melding, is the alchemy of love. For those who are ready, vision is amply available in that space. It is hard wired into the architecture of the cathedral, all the ingredients set into the crucible. I have heard rabbis speak this way of the Torah or the Kabbala; and I have heard poets speak this way of *The Divine Comedy*. At Saint Francis' basilica, the effect is extraordinary. There is a palpable rhythm. The lower church is designed as a tomb, the ceiling lower—"crushed" architecturally—as though the upper church were collapsing into it. It's a mortar with the pestle pressing down, and you are the powdery substance it's crushing. It offers an unconscious illusion of weight, of the sky falling. Immediately you feel claustrophobic, like being buried alive. The most famous fresco down below is Pietro Lorenzetti's *Madonna With Child, Saint Francis and Saint John the Baptist*. It's really rather small compared to other works by the Giotto school, and Cimabue and Martini. In it, Mary is holding Jesus and points towards Francis. She is pointing with her thumb; almost casually, or secretly. To her right is St. John looking shocked. Mary is endorsing Francis over John. John is holding his bible and looking left out. Francis is pointing to himself, shocked too, a "Who, me?" expression crossing his face. Near the right transept is Cimabue's *Madonna in Majesty*, significant because in it is a portrait of Francis that is said to be the most accurate we have; the details of his likeness had come from Francis' own nephews. He is hunched and his ears stick out at right angles; his eyes are droopy and huge. He bears the look of someone very ordinary. The place on which the basilica stands is called "Paradise Hill," though it had once been called "Hell Hill," Colle d'Inferno, legend claims. People had been executed there in the years before Francis' birth. The image of life emerging from a dead thing pervades planting cultures the world over. From the dead tree new shoots appear. It was an observation gleaned from nature, but applied to spiritual life. Jesus is crucified on the "place of skull" where Adam was said to be buried. From the fallen comes the risen.

Below the basilica is the unearthed tomb of the man himself. Crowds of people, mysteriously, do not interrupt the aura of sanctity in this place. It is, it seemed to me when I visited the first time, the holiest place I had ever been. Arranged directly in front of the tomb are about five or six rows of pews, where sit nuns and priests and pilgrims from every worldview and culture. I have to say, I couldn't stay there long. Something about that place really spooked me. Think of the first time you saw a real corpse. Smaller tombs surrounded him, friar-friends of the saint. I could only stand it for ten minutes, no more, and by then I felt suffocated. I left by the basement door, which led into an open courtyard overlooking the olive fields of Umbria; it was bright, and I hiked my way up to the street.

Notes

1. Brassard, Nigel à, *Keats in Rome* (London: The Letter Press, 2003).

2. Levine, Philip, *So Ask: Essays, Conversations, and Interviews* (Ann Arbor, MI: University of Michigan Press, 2005).

3. Keplinger, David, in *The Clearing* (Kalamazoo: New Issues Poetry and Prose, 2005).

4. Eliot, T.S., *The Sacred Wood* (London: Methuen & Co, Ltd., 1960).

5. Alighieri, Dante, *La Vita Nuova*, Trans. Barbara Reynolds (Baltimore: Penguin Books, 1960).

6. James, William, *The Varieties of Religious Experience* (New York: Penguin Books, 1982).

Alison MacLeod

Alison MacLeod *grew up in Canada and moved to the UK in 1987 after accepting a place on the University of Lancaster's MA program in Creative Writing. She published her first short story that year and has since gone on to publish stories in a variety of UK and Canadian magazines and collections. Her first novel,* The Changeling, *was published by Macmillan UK and St Martin's Press in 1996. Her second novel,* The Wave Theory of Angels, *was published by Hamish Hamilton and Penguin Canada in 2005. Her short story collection,* Fifteen Modern Tales of Attraction, *will be released by Penguin UK in the autumn of 2007. She teaches on the MA in Creative Writing at the University of Chichester and lives in Brighton.*

MAKING LITERATURE, SCIENCE—AND LOVE?

ALISON MACLEOD

I. Strange Bedfellows

> [T]he first criterion of success in any human activity, the necessary preliminary, whether to scientific discovery or artistic vision, is intensity of attention, or, less pompously, love.
>
> W. H. Auden[1]

It's a provocative claim. Indeed, one might wonder why Auden, writing in *The Criterion* in 1933, didn't confine himself to that seemingly more appropriate phrase, "intensity of attention", or even "fixity of purpose" as he characterizes it later in the piece, for which artist or scientist wouldn't concede the role of focus and determination in any breakthrough endeavour?

But love? "The *first* [my emphasis] criterion of success" in "scientific discovery or artistic vision"? Auden does not elaborate. He's unabashed. The word "love" dangles like a showy upstart.

I'd like to explore at least something of that which Auden has left unsaid. How might love catalyse processes of creation, and specifically literary creation, my own area of interest, as well as scientific discovery? Could love really be *the* vital dynamic behind worlds as apparently disparate as art and science? Is love, along with those more familiar qualities of focus and determination, crucial to far seeing vision?

In 1969, the iconic English novelist John Fowles described an image that "rose in his mind" when he was "still in bed half asleep".[2]

> It corresponded to no actual incident in my life (or in art) that I can recall . . . These mythopoeic 'stills' (they seem almost always static) float into my mind very often. I ignore them, since that is the best way of finding whether they really are the door into a new world.
>
> So I ignored this image; but it recurred. I began deliberately to recall it and to try to analyse and hypothesize why it held some sort of imminent power. It was obviously mysterious. It was vaguely romantic. It also seemed, perhaps because of the latter quality, not to belong to today. The woman obstinately refused to stare out of the window of an airport lounge; it had to be this ancient quay—as I happen to live near one, so near that I can see it from the bottom of my garden, it soon became a specific ancient quay. The woman had no face, no particular degree of sexuality. But she was Victorian; and since I always saw her in the same static long shot, with her back turned, she represented a reproach on the Victorian age. An outcast. I didn't know her crime, but I wished to protect her. That is, I began to fall in love with her.[3]

The woman was Sarah, the eponymous heroine of Fowles' most acclaimed novel, *The French Lieutenant's Woman*.

Reading the above, an apparent declaration of love for a fictional character, one would assume Fowles is merely employing a stock turn-of-phrase to make a point. He was, we understand, struck by a visual image. It was an inspirational moment, even a gift of a moment, but a moment that, in reality, had little to do with love. How could it? Perhaps we intuit that, even here in an essay about the making of a novel, the novelist cannot resist the dramas of story-telling, and we gladly forgive the fault, if fault it is. Later, he confesses, "I loathe the day a manuscript is sent to the publisher, because on that day the people one has loved die . . ."[4]

But Fowles is not playing fast and loose with words. "Characters (and even situations)," he explains, "are like children or lovers, they need constant caressing, concern, listening to, watching, admiring. All these occupations become tiring for the active partner—the writer—and only something akin to love can provide the energy."[5]

At some level then, the character of Sarah is "alive" as Fowles writes her, and, living being that she is, she makes demands:

> I was stuck this morning to find a good answer from Sarah at the climax of a scene. Characters sometimes reject all the possibilities one offers. They say in effect: *I would never say or do a thing like that.* But they

> don't say what they would say; and one has to proceed negatively, by a very tedious coaxing kind of trial and error. After an hour over this one wretched sentence, I realized that she had in fact been telling me what to do: silence from her was better than any line she might have said.[6]

So "love" for Fowles is not merely dedication to a project. It is *devotion* to a character or an imaginative situation that has arrived seemingly unbidden, an invaluable offering of the creative process. And in this devotion, as in all forms of devotion, there is an element of personal sacrifice, an almost literal giving-over of one's self to the "life" of that creation. In the above account, it is indeed difficult to say where Fowles ends and Sarah begins. "All natural or born writers," Fowles writes, "are possessed, and in the old magical sense, by their own imaginations . . ."[7]

The comparison of authorship to the state of possession is not, for Fowles, a casual one, nor is it without precedent. "Madame Bovary, c'est moi," Flaubert once famously and cryptically declared. Charlotte Brontë was more forthcoming. In the preface to the second edition (1850) of her late sister Emily's novel, *Wuthering Heights*, she offers a glimpse into the creation of the elemental, degenerate and now legendary character Heathcliff. Her remarks are a lightly veiled attempt to defend Emily Brontë and the novel against its numerous critics who objected to the characterisation of Heathcliff on moral grounds.

> Whether it is right or advisable to create beings like Heathcliff, I do not know: I scarcely think it is. But this I know: the writer who possesses the creative gift owns something of which he is not always master—something that, at times, strangely wills and works for itself. He may lay down rules and devise principles, and to rules and principles it will perhaps for years lie in subjection; and then, haply without any warning of revolt, there comes a time when it will no longer consent to "harrow the valleys, or be bound with a band in the furrow" Be the work grim or glorious, dread or divine, you have little choice left but quiescent adoption. As for you—the nominal artist—your share in it has been to work passively under dictates you neither delivered nor could question—that would not be uttered at your prayer, nor suppressed nor changed at your caprice. If the result be attractive, the World will praise you, who little deserve praise; if it be repulsive, the same World will blame you, who almost as little deserve blame.[8]

What Charlotte cannot say to the indignant reviewers of *Wuthering Heights* is that, for all his diabolical rage and evil-doing, Heathcliff is a compelling character, compelling not because he is monstrous, but because he is contradictory and complex. Grotesque in his demands, he is also utterly human, blighted by loneliness and a gnarled longing for intimacy. Such subtlety of characterisation, I would argue, can only have been achieved through a constancy of attention—or, in Fowles' words through "something akin to love".

Re-reading Brontë's words ("As for you—the nominal artist—your share in it has been to work passively under dictates you neither delivered nor could question . . ."), I am reminded of Fowles the novelist as he is instructed by Sarah the character in what she will and will not say. I am reminded, too, of the contemporary novelist, D.M. Thomas's comment about his character Lisa Erdman, the heroine of his groundbreaking 1981 novel, *The White Hotel*: '"I kept discovering new things about Lisa Erdman. Much that was in her letters [in the novel] to [the character of] Freud took me by surprise."[9]

Who then is in control, author or character? "A writer doesn't choose his subject matter," argues D.M. Thomas, "he submits to it".[10] How, after all, could a vicar's daughter leading a sheltered life in remote nineteenth-century Haworth "choose" or decide upon the character of Heathcliff? The question is commonplace in literary-critical circles. In those same circles, of course, the notion of "literary possession" is occult, in the sense that it is literally occluded from active critical discussion. In my experience, the only critic who explores this uncanny but vital dynamic of literary creation in general is the great Russian thinker, Mikhail Bakhtin.[11]

Perhaps the truth of Heathcliff's conception lies in the conclusion to Charlotte Brontë's preface. Here, in a passage that reads too passionately to be merely figurative, Charlotte evokes Emily's keenly felt sense of the "spirit of the place"—the place in this case being the wild moorland that, even today, surrounds the Yorkshire village of Haworth, a landscape to which Emily, of all the Brontë children, was most devoted. "As far as the scenery and locality are concerned, it could scarcely have been so sympathetic: [she] did not describe as one whose eye and taste alone found pleasure in the prospect; her native hills were far more than a spectacle; they were what she lived in, and by . . ." And in Emily's "hewing" of the character of Heathcliff, the spirit of the place is made flesh: "With time and labour, the crag took human shape; and there it stands colossal, dark, and frowning, half statue, half rock; in the former sense, terrible and goblin-like; in the latter, almost beautiful, for its colouring is of mellow grey, and moorland moss clothes it..."[12]

Indeed, *could* the character of Heathcliff in some way be the manifestation of the spirit of that moorland? *Could* Emily Brontë, in her love for her native moorland, have been the devotee and medium through which the land "spoke" or "thought" itself?[13] Could she have, in some sense, *become* the spirit of that landscape as she wrote?

In a daring and poignant exploration of the life of the poet Walt Whitman, the writer Lewis Hyde makes a convincing connection between artistic creation, the state of possession—or "enthusiasm" as he terms it—and love:

> To be "enthusiastic", originally meant to be possessed by a god or inspired by a divine afflatus. The bacchants and maenads were enthusiasts, as were the prophets in the Old Testament, the apostles of the New, or, more recently, Shakers and Pentecostal Christians. Enthusiasts, having received a spirit into the body, have never been hesitant to speak of a "sweet burning in the heart" or a "ravished soul".[14]

For Hyde, certain gifted writers or artists are also enthusiasts.

No one can speak for Emily Brontë, or testify to her precise experience as she wrote her novel, but certainly *Wuthering Heights* is aflame with that "sweet burning in the heart". "I *am* Heathcliff," declares its heroine, Catherine Earnshaw, "—he's always, always in my mind—not as a pleasure, any more than I am always a pleasure to myself—but as my own being."[15] As a writer, I cannot help but wonder if Cathy speaks not only for herself but also for her maker; if Emily Brontë was also and irrefutably Heathcliff; if she was an enthusiast of that ancient variety, possessed and impassioned by the *genius loci*, or animating spirit of the moorland she loved. Are the realities of creation more complex than we can usually articulate? In the words of the African-American novelist Toni Morrison, "Imagining is not merely looking, or looking at; nor is it taking oneself intact into the other. It is, for the purposes of the work, becoming."[16]

II. Science and the Unbidden

It is possible that, in culture at large, we are not very surprised that artists seem privy to such mystery. Indeed, post-Blake, Coleridge and Wordsworth, perhaps we even expect a certain other-worldliness of our first-rate talent. It allows us to enjoy, if briefly and superficially, a very literal *je-ne-sais-quoi*. However, if we return again to W.H. Auden's risky claim, scientists, we generally agree, operate very differently. Toni Morrison's "becoming", Charlotte Brontë's "possession", Lewis Hyde's "enthusiasm", and John Fowles' "love" simply don't come into it.

Or do they?

In 2001, I listened to an interview with Cambridge scientist John Sulston on BBC Radio 4's *Desert Island Discs*. Sulston was, of course, one of the leading figures in the recent turn-of-the-century race to map the human genome. Throughout the interview, his dedication to the project was clear. So too was what emerged as an early passion for science and a keen childhood curiosity about how the physical world worked. Yet it was a particular choice of words at one point, a particular emphasis, that struck me, in his description of a boyhood moment: "I remember trying and trying to figure out how a transformer worked, how it changed electricity; converted it from here to here . . ." Standing in my Sunday-morning kitchen beside my radio and fruit bowl, I was fascinated as something unexpected and strange unfolded in Sulston's account: "I looked and looked," he explained, "till I *became* the transformer, till I could *feel* the electricity, the voltage, the magnetism."

"[T]ill I *became* the transformer . . ." He had chosen his words carefully. The moment was both long-ago and alive again as he described it. Somehow, under the rapt intensity of his boyhood gaze, in that flash of sudden knowledge, he and the transformer had apparently become one, even as Fowles and Sarah had become one, or D.M. Thomas and Lisa, or Emily Brontë and Heathcliff. At the age of 59, Sulston sounded like a boy who'd fallen in love again for the first time.[17]

In *The Gift: Imagination and the Erotic Life of Property*, Lewis Hyde reminds us that *eros* is, first of all, the principle "of attraction, union [and] involvement which binds together" and that it exists in opposition to *logos*, or "reason and logic in general, the principle of differentiation in particular".[18] In 21st-century Western culture, we tend of course to be more at ease under the directives of *logos*, and who would be without it? Which genuine artist or scientist could spare this faculty which allows us to reason, to make distinctions, and to see the beautiful particular in the general? We are less comfortable, however, with the force of *eros*, that live current not only of human love, but, as Hyde argues, of wholeness too.

Is it then in the keenly desired union with something *other* than ourselves (the "mythopoeic still" of Sarah, the Yorkshire moorland, the electrical transformer) that we experience that rare sense of unbidden wholeness? Is it only through this "erotic" coupling of the passionate self and an other that we stumble upon the gift of holistic awareness?

John Milton wrote that the Muse "dictated" to him the whole "unpremeditated song" he would later entitle *Paradise Lost*. Samuel Taylor Coleridge famously woke from sleep with what he described as "a distant recollection" of the whole of the poem "Kubla Kahn". Nor is this experience of "wholeness" and certainty exclusive to artists. The mathematician Henri Poincaré recalled one such momentous experience:

> The changes of travel made me forget my mathematical work. Having reached Coutances, we entered an omnibus to go to some place or another. At the moment when I put my foot on the step, the idea came to me *without anything in my former thoughts seeming to have paved the way for it* [my emphasis], that the transformations I had used to define Fuchsian functions were identical with those of non-Euclidean geometry. I did not verify the idea; I should not have had the time, as, upon taking my seat in the omnibus, I went on with a conversation already commenced, but I felt a perfect certainty.[19]

And he wasn't wrong. In 1914, Poincaré made the breakthrough discovery of Fuchian functions in mathematics.

Coincidentally, Friedrich Kekulé's fundamental theory of organic molecular structure also, famously, arrived on a London bus, the last night bus out of Islington:

> I fell into a reverie, and lo, the atoms were gamboling before my eyes! Whenever, hitherto, these diminutive beings had appeared to me, they had always been in motion; but up to that time I had never been able to discern the nature of their motion. Now, however, I saw how, frequently, two smaller atoms united to form a pair; how a larger one embraced the two smaller ones; how still larger ones kept hold of three or even four of

> the smaller, whilst the whole kept whirling in a giddy dance. I saw how the larger ones formed a chain, dragging the smaller ones after them but only at the ends of the chain . . . The cry of the conductor: "Clapham Road," awakened me from my dreaming . . .[20]

Kekulé awoke from his dream of this "giddy dance" with the explanation of how carbon atoms link together to form chains, an explanation that would inspire his 1858 *Theory of Molecular Structure*.

Yet he was also aware that he'd failed to account for the whole of organic chemistry. For the next seven years he would labour over the puzzle of "aromatic compounds", trying to produce the structural formula that would account for their distinctive character. Then, one evening in Ghent, he dozed by a fire.

> . . . I resided in elegant bachelor quarters in the main thoroughfare. My study, however, faced a narrow side-alley and no daylight penetrated it. For the chemist who spends his day in the laboratory this mattered little. I was sitting writing at my textbook but the work did not progress; my thoughts were elsewhere. I turned my chair to the fire and dozed. Again the atoms were gamboling before my eyes. This time the smaller groups kept modestly in the background. My mental eye, rendered more acute by repeated visions of the kind, could now distinguish larger structures of manifold conformation: long rows, sometimes more closely fitted together, all twining and twisting in snake-like motion. But look! What was that? One of the snakes had seized its own tail, and the form whirled mockingly before my eyes. As if by a flash of lightning I awoke; and this time also I spent the rest of the night working out the consequences of the hypothesis.[21]

Kekulé's insight that carbon atoms form rings as well as chains was published in 1866. As John Carey notes, nature is based on the carbon chain and the carbon ring, and "life itself relies on the capacity of carbon atoms to form molecular chains and rings as they did in Kekulé's dreams."[22]

In 1921, the states of sleep and waking were also crucial for the German-American pharmacologist Otto Loewi. It would lead to his discovery that the transmission of nerve impulses to organs is chemical, not electrical.

> The night before Easter Sunday . . . I awoke, turned on the light, and jotted down a few notes on a tiny slip of paper. Then I fell asleep again. It occurred to me at 6 o'clock in the morning that during the night I had written down something most important, but I was unable to decipher

> the scrawl. The next night, at 3 o'clock, the idea returned. It was the design of an experiment to determine whether or not the hypothesis of chemical transmission [of the nervous impulse from nerves to their respective organs] that I had uttered 17 years ago was correct. I got up immediately, went to the laboratory, and performed a simple experiment on a frog heart according to the nocturnal design.[23]

In 1936, Loewi was awarded a Nobel Prize for his discovery of the chemical transmitter acetylcholine, a discovery that would have an impact "on everything from our knowledge of brain function to the treatment of neurological disease".[24]

"There are no logical paths to such natural laws," Einstein would say of his own discovery of Special Relativity, "only intuition can reach them." The "happiest thought" of his life, he recollected, was a mental vision of a person falling off a roof in which came also the whole and sudden understanding that someone in this position was at rest *and* in motion at the same time.[25] There is now a small sign at a window in Einstein's apartment through which one can look; the sign explains that through this very window, special relativity was born.

Likewise, string theorist Michio Kaku remarks, "I find myself spending most of my time staring out the window. I see blocks of equations dancing in my head, and I spend hours trying to fit them together."[26] A window is a dream-space, as was Loewi's bed and Kekulé's hearth. For the scientist too, dreams and mental visions are the spaces through which that force of sudden wholeness, love or *eros*, can enter, bringing with it living forms—Kaku's dancing equations, Loewi's fervent jottings, Kekulé's gamboling atoms.

Yet the notion of *eros* and "living forms" has no place of course in the Scientific Method. How can it? Its time-honoured standard procedures and criteria demand consistency, predictability and reproducibility. The brief testimonies of Sulston, Poincaré, Kekulé, Loewi and Einstein can only exist outside it as curiosities or anecdotal anomalies in a scientific tradition that first distinguished itself from centuries of superstition by spurning all claims to knowledge that were not directly attributable to recorded observation.

The Scientific Method is not the problem. Its success rate, across scientific disciplines, speaks powerfully for itself. But perhaps we need alternative questions and tools with which we can approach and, most importantly, engage with the physical world. For Wolfgang Pauli, one of the most respected physicists of the last century, the burning issue was "the lack of soul in the modern conception of the [physical] world". For "soul", we might also read "animating spirit". As physicist F. David Peat relates in his article, "Wolfgang Pauli: Resurrection of Spirit in the World", Pauli despaired, in both his personal and professional life, that "he continued to live in a world where there was a clear split between spirit and matter".

As Peat explains, Pauli was visited in his dreams on several occasions by a Persian stranger, a figure whom he believed "was tempting him to enter the world of the senses". If so, it would be fitting guidance, for is it not through a deep engagement with the sensual that reality comes most fully and vividly to life, as in John Sulston's early experience of looking and looking at that transformer in all its transformer-ness? Significantly, in a letter to Carl Jung, who was counselling Pauli at the time, Pauli came to the painful conclusion "that the missing element was Eros".[27]

Lewis Hyde would not be surprised.

But what does it mean to enter the world of the senses? The apprenticeship of the scientist and the poet is, in some ways, much the same; both disciplines demand that its initiates train themselves to be keen-eyed observers of the physical world. But potentially, it is in this most obvious and, in a sense, everyday of places that we experience what the writer James Joyce called "epiphany"; where we can grasp, as he put it, "the what-ness of the thing". Yet is it also at this junction between revelation and imaginative understanding that the poet and scientist must finally part ways?

Niels Bohr, father of quantum theory, didn't seem to think so.

> "We must be clear that, when it comes to atoms, language can be used only as in poetry," he told Heisenberg one day as they trekked through the German woods. "The poet, too, is not nearly so concerned with describing facts as with creating images and establishing mental connections." Then, Heisenberg asked, "how can we ever hope to understand atoms?"
>
> "I think we may yet be able to do so," Bohr replied. "But in the process we may have to learn what the word 'understanding' really means."[28]

What *is* the essence of poetic understanding, and what might it reveal to the scientist and the layperson about the physical world?

The "atom" of poetic reality is the metaphor. A metaphor of course is a "conjugation" of two seemingly disparate images from which a fresh and undreamed-of "third" is born. It is naturally "conjugal"; so, it's a marriage of sorts or a coupling.

Ask yourself. What could a fallen apple possibly have to do with a slant of sunshine? Recall Dylan Thomas' effortless image of "the windfall light". It seems at once surprising and inevitable. Yes, we say. Of course.

Metaphor allows us to break with habitual or received ways of seeing and knowing the world. What could a potato have to do with the moon? Yet the two become one as Thomas pulls back the curtain of the night to point to "the halfmoon's vegetable eye". In this new coupling, both potato and moon become, simultaneously, more vivid *and* more strange to us. They become, in other words, fuller, more striking versions of themselves.

Even in these very simple examples, one rides the life-force of language, and the drive toward connection is essentially erotic. Words unhitch themselves from fixed meaning; they go forth and multiply. Language (our categories, our classifications) slips and shifts, and with it, so too does perception. In the process, the world itself becomes seemingly mutable and abundant; something that exceeds our ability to define it; something bigger than we are. Something, in other words, that is alive.

We can feel the dizzying and lovely flux of the so-called ordinary world in the American poet Mark Doty's poem, "Difference". Here, in a poem about the common jellyfish, he suggests that, in fact, *all* seeing is always "seeing like" or "seeing as"; that all perception is always innately and helplessly metaphorical, because the physical world can never, at its most essential, be fixed; because it is always stubbornly myriad.

DIFFERENCE

The Jellyfish
float in the bay shallows
like schools of clouds,

a dozen identical—is it right
to call them creatures,
these elaborate sacks

of nothing? All they seem
is shape, and shifting,
and though a whole troop

of undulant cousins
go about their business
within a single wave's span,

every one does something unlike:
this one a balloon
open on both ends

but swollen to its full expanse,
this one a breathing heart,
this a pulsing flower.

This one a rolled condom,
or a plastic purse swallowing itself,
that one a Tiffany shade,

this a troubled parasol.
This submarine opera's
all subterfuge and disguise,

its plot a fabulous tangle
of hiding and recognition:
nothing but trope,

nothing but something
forming itself into figures
then refiguring,

sheer ectoplasm
recognisable only as the stuff
of metaphor. What can words do

but link what we know
to what we don't,
and so form a shape?

Which shrinks or swells,
configures or collapses, blooms
even as it is described

into some unlikely
marine chiffon:
a gown for Isadora?

Nothing but style.
What binds
one shape to another

also sets them apart
—but what's lovelier
than the shapeshifting

transparence of *like* and *as*:
clear, undulant words?
We look at alien grace,

unfettered
by any fixed form,
and we say: balloon, flower,

heart, condom, opera,
lampshade, parasol, ballet.
Hear how the mouth,

so full
of longing for the world,
changes its shape?[29]

The "its shape" of the final line is ambiguous. Does "its" refer to the mouth changing shape in its longing to utter the world? Does the mouth become as mutable as the jellyfish it describes? For, paradoxically, isn't that mouth (and by extension, the poet) *also* part of the living flux of the created world it captures? Can anything ever stand outside it?

Or does "its" refer to "the world"? Is Doty suggesting that the physical world itself changes shape through the force of the poet's longing; through the force of his desire to hold it, fleetingly, with language? Does desire somehow catalyse a Heraclitian "everything flows" transformation? Does it lay bare, or even engender, a flux that is perhaps the fundament of the physical world?

I suspect all of the above is possible. I also suspect that Niels Bohr, were he here, might be inclined to agree; that he, like Doty, understood the rich potential of images (those born of sharp observation) to capture reality at its widest and most unpredictable; that he, too, understood that words are, in fact, only marvellous images, and that tried-and-true facts, at some point, always become provisional. In Doty's poem, the living "what-ness" of the jellyfish is revealed in a slipstream of images that are both recognisable and fresh; everyday ("balloon", "flower", "heart", "condom") and urgent; unassuming and incantatory.

Stanza by stanza, the poem rises to a final revelation of desire and longing, and it is this force of love, I would argue, that has allowed the poet to bring back from the inchoate sea of jellyfish facts and observations the *life* of his subject, and, with it, by implication, a sense of the full life of the wider physical world of which the jellyfish are but one detail. The metamorphosis of the final lines, in which both the poet's mouth and the world at large simultaneously begin to shape-shift, takes place not on the page itself, but finally in the mind of the reader. We too enter the slipstream, and become one with the poet, his mouth, the jellyfish, the purse, the lampshade and the condom. In this "all's one" reality made accessible to us by the poet, the world is met and held, *alive* in our own mind's eye.

Perhaps, for these reasons, the scientist, too, needs literature and art, not just as an outlet from the demands of *logos*, but as a bridge, a point of contact, to a wider experience of the real. Perhaps every scientist should also learn, firsthand, something of the craft of the poet.

III. On *The Wave Theory of Angels*

To be honest, I'd never given much thought to angels. If anything, outside of certain artistic traditions, I found the subject faintly embarrassing, the stuff these days of cheesy Christmas calendars and television phone-ins.

Wim Wenders' 1987 film, *Wings of Desire*, had long been an exception. Here, two male angels dressed in sober black overcoats, walk (and occasionally hover over) the streets of a war-scarred Berlin, until one of them falls helplesly in love with a circus girl and longs to be human. He yearns to leave footprints in the snow, to crunch on an apple, to see red, to make love; to be more, in short, than a divine abstraction. As he stumbles into mortality, the film flickers from black-and-white into Technicolor, and we see our world through his fresh eyes. The cityscape of Berlin unfolds, and we remember all over again how distinctive the ordinary world is, even in the dirt of a broken post-war city; even in the spit and sawdust of a bankrupt circus. Indeed, the dirt, spit and sawdust are also somehow illogically beautiful.[30]

In Wenders' film, the angel's physical transformation is born of his longing. His desire catapults him into mortality. Yet the mythic (or psychical) connection between desire and transformation is age-old. In Ovid's *Metamorphoses*, as the poet Ted Hughes once remarked, mortals and gods spontaneously combust, mutate or levitate under the terrible pressure of their own desires.[31] Hughes was right. As readers, we can only watch in wonder, amusement and sympathy as their assorted longings bear down on them with an emotional g-force the physical world itself cannot resist.

The Wave Theory of Angels began, long before it began, when I stumbled upon and fell in love with a very brief account of an apocryphal medieval saint. Christina the Astonishing died while still a young woman, and, in a startling reversal of all expectations, sat up on the bier during her own funeral mass. When she soared to the rafters of the church, the congregation, understandably, fled. Only her sister stayed and tried to talk her down.

The Church had long ago wiped her from its calendar of feast days, but that only recommended her to me all the more. She was transgressive. I loved the fantastic, Ovidian quality of Christina's flight from death. I loved the image of that sudden and defiant surge of life. What desire, I asked myself, could have fuelled so powerful a change? What longing could have catalysed such a return to life?

I wanted to create a story and a space in which a modern reader might suspend disbelief just long enough to wonder—in both senses of that word; i.e., to marvel and to ask questions. I wanted to evoke a fleeting sense that, even in the 21st century, reality still resists our efforts to pin it down. Ours is the Information Age, and at times there has seemed little that rational enterprise cannot know or accomplish. Indeed, were it not for the horror of our most rationally executed disasters (the Final Solution, Hiroshima, Nagasaki, Cambodia, Sept.11th), the twentieth and twenty-first centuries might even have been mistaken for the flowering of the eighteenth-century Enlightenment Project. But imbalance is invariably a dangerous thing.

In highly rational times, we necessarily sacrifice those experiences of the 'irrational' that renew us: that's to say, our ability as a culture to wonder, to marvel, to receive mystery, and to make leaps of faith or intuition. A small price, many would argue. Yet the upshot is, that without it, we, arguably, make ourselves vulnerable to the dark flip-side of the irrational: that's to say, to superstition, to religious hatred and to brute fundamentalism—radical Islam's fervour, Bush's war-on-terror "Crusade", and Blair's zealously moral high ground.

The Wave Theory of Angels unfolds in the years 1284 and 2001. Both the medieval and the modern-day storylines explore this dark potential of the irrational that now, tragically, seems as much a hallmark of the 21st century as it was of the 13th. In fact the novel was already, almost independently of me, running toward these questions, when, all too terribly out of the blue, Flight 11 and Flight 175 hit the World Trade Center.

In August 2000, immersed in my research into13th-century life, culture and thought, I came across a phrase that struck me as interesting if tangential at best: 'Fear of the Angel'. My source referred very briefly to a controversy that had nearly brought down the University of Paris in the early part of the 13th century. The work of the Islamic (or so-called "Arabist") philosopher Avicenna had suddenly been banned. I knew nothing of his work, but I chased up a title, *Avicenna and the Visionary Recital*,[32] quickly added it to a list of some twenty seemingly more pertinent titles, and forgot about it.

That night, I did what I sometimes do; on the brink of sleep I, rather madly and not a little desperately, asked my dreaming mind for help with my still very tentative novel. If I'm honest, it's not quite as mad as it sounds. Sometimes, I seem able to direct myself toward useful information or insights in dreams that will conveniently wake me with the sharp nudge of their own relevance. At this point in the process, I was certainly feeling stuck. I had a few vital elements for the novel, but I had little idea of how to make the story "go".

That night, however, the experience was both decisive and, for me, profound.

> I'm at Yaffles, though the house is very different. Jill tells me she is getting rid of lots of books; she has stacked them in the corridor. Do I want any?
>
> The upstairs corridor is very long and narrow, unlike the actual corridor

> I remember well. It's really more of a passageway. I say I will look at the books, really only to be polite. I lift one from the stacks. As I do, its cover starts to glow. I look more closely. It's covered in wings. As I look they move and fill with light. I realise it's an angel on the front. I look at the title: Avicenna's Visionary Recital. It's pure pleasure to see it, to hold it. "It's so funny," I say to Jill, "because I ordered this through inter-library loan only today."

When the books arrived, it was the first I turned to. Perhaps the dream's alteration of the wide corridor of the family home I had once known into something I could identify only as a "passageway" was significant; for it was to emerge as perhaps *the* vital passageway through what had been a very tight spot in my writing process. The controversy known as "Fear of the Angel" offered me the way forward.

Once upon a time, the Christian West had taken its cues from Islamic scholars. Avicenna was an 11th-century physician, philosopher, and a respected authority on Aristotelian logic, metaphysics, psychology and the natural sciences. Yet, in the 13th century, the question of Avicenna's "tenth angel" would bring uproar to the University of Paris.

Suddenly, nine archangels were deemed sufficient; a tally of nine nine meant there was one angel to oversee each heavenly sphere. Avicenna, however, counted ten, and his tenth angel was not celestially remote. On the contrary, this angel occupied the sphere between the heavens and our earth, and, more surprising still, he was open to union with each mortal. Human longing was the catalyst. The plain of the imagination was the meeting place.[33] The beloved detail of the ordinary physical world was the shared language. Communication and revelation took the form of images, not speech. Through this loving union, the physical detail of Creation was renewed. Through this imaginative collaboration, Creation also evolved.

So the tenth angel was trouble. Not only was he awash in all that desire, not only did the concomitant notion of an evolving universe threaten the status quo, but the tenth angel also meant that we ordinary humans could participate *directly* in the divine cosmos; that the Church's intercession was not strictly necessary.

I realised that Christina's father—the "Imaginator", or sculptor, Giles of Beauvais—was who he was, above all because he stood for the potential of the imagination. As my novel took life again, I knew I had to locate him at the University of Paris in his youth; that he'd been a student who had argued passionately for the importance of the tenth angel. For Giles, the stakes would have been high, for in those days, metaphysics was *also* physics; angels were also the very atoms of the universe, and physics was a reading of the mind of God. Giles, then a student philosopher arguing against the ban on the "Arabist" philosopher, could only have been branded a heretic and cast out. Indeed, he would have been lucky to escape Paris with his life.

Ultimately, Avicenna's tenth angel is perhaps nothing more (and nothing less) than an embodiment of the principle of Eros, that force of connection and wholeness in the world. Angel or principle, we seem to be talking about a live force in the world, odd as that, admittedly, does sound. Yet with Fowles' loving creation of the "living" character of Sarah, or Brontë's creation of the strangely wilful Heathcliff, in Doty's rapt rendering of the shifting jellyfish, and in Sulston's own transformation through the transformer, perhaps we *do* actually see the world evolve, detail by surprising detail.

I recorded two such moments in my journal, moments in which "reality" seemed transfigured. Each was brief, no more than a few minutes, yet each was also quietly extraordinary. The first came in April 2000, a few months before my discovery of the tenth-angel controversy.

> I wake and see above my head, a small flying thing, hovering very bright in the darkness. It is hard to name at first because it is so much in motion. Like a butterfly, only the size of a small bird. It glows red and blue, as if made of neon or a naked flame. I try to focus. I can make out an electric blue body and wings, with edges of red . . . A quick vision but much more than a dream. I have the sense that I am actually watching a living thing that I might easily have reached out and touched had I dared.

The second came some six months later:

> A night vision again. I am awake, eyes open to the night; am watching a huge pair of wings, sometimes black and sometimes blue, flapping high overhead, some 10 feet up, near the top of my big Edwardian wardrobe at the far end of my room. I can only stare, mesmerised and a little frightened, until they disappear.

Looking back, it seems only natural that I should have been "visited" by wings, or at least by the image or vision of "winged-ness"; the force of "uplift", both liberating and dark, would be vital to *The Wave Theory*.

Yet involved as I was in the tenth-angel "Arabist" controversy, I couldn't rest in the medieval world alone. I was writing for a modern readership, and while the 13th-century setting and characters provided a very natural launch for questions about imagination and belief, I was loathe to leave my story there. I was concerned it would be too easy to dismiss images of mystery as the by-products of medieval ignorance and spectacle. I needed to transplant these images and questions to the here and now. What sort of modern-day world could provide the vital backdrop? Where could the story of Christina and Giles find a home in the year 2001? Anything that smacked of New Age mysticism was out. I needed a world that would seem reasonable and credible to the reader.

I had just finished reading Danah Zohar's *The Quantum Self* (1991). Over the next five years, I would come to love quantum theory's reluctant notions of indeterminacy, uncertainty, complementarity, non-locality and entanglement. I would immerse myself in the philosophers of the new physics. I would book myself on courses, and chase after an understanding of String Theory and M-Theory. I would fly to Chicago to visit Fermilab, the renowned physics laboratory—and the workplace, I would soon realise, of my 21st-century character, Giles Carver. Like his medieval counterpart, Dr. Giles Carver would also wrestle with the mysteries of the invisible world, even as the mystery of his sleeping daughter would frustrate his abilities to explain it.

Once again, here at the quantum frontier, was the world in flux; here, too, the boundaries between dimensions seemed to give way; here, too, there was the ghost of a suggestion that we might co-create the world; that ours, like Avicenna's, might be a participatory universe. For this reason, the Observer Effect of quantum theory was hard to resist. Here, human consciousness is posited as *the* force that collapses quantum wave function into the fixed and real; in other words, the force that sieves physical reality from the flux of sub-atomic possibilities. The idea excited me, verifiable, accurate or not. The debate was enough— great novels have been based on far less. If human consciousness could act on the physical world, by implication, so could the imagination, including the reader's own.

From page 72, the reader is him- or herself invoked, as a witness to the novel's events, and more specifically, as the catalytic Observer of the

Observer Effect. Of course, the reader is always involved in the creation of any novel, but with *The Wave Theory*, I wanted the power of the imagination to feel real, active, firsthand, and not merely a literary given. I wanted to create an experience of the novel as something alive and unfolding. I wanted to involve the reader directly in its page-by-page evolution; to make it somehow co-creative. I wanted the reader to want.

It was only at this point—desiring more for the sake of my novel than I could perhaps deliver—that I returned to page one, knowing what the first lines must now be:

> The world yearns. This is its sure gravity: the attraction of bodies. Earth for molten star. Moon for earth. A hand for the orb of a breast. This is its movement too: the motion of desire, of a longing toward.

Or, as even Auden might have it, the motion of love.

Notes

1. W.H. Auden/Edward Mendelson (ed), *The English Auden: Poems, Essays and Dramatic Writing, 1927-1939* (Faber, 1988), p. 319.
2. John Fowles/Malcolm Bradbury (ed), 'Notes on an Unfinished Novel', *The Novel Today: Contemporary Writers on Modern Fiction* (Fontana, 1977), p. 147.
3. ibid, pp. 147-48
4. ibid, p. 160
5. ibid, p. 148
6. 'Notes on an Unfinished Novel', pp. 158-59
7. ibid, p. 148
8. Charlotte Brontë (ed)/ Emily Brontë, 'Editor's Preface to the New [1850] Edition of *Wuthering Heights*', *Wuthering Heights* (Penguin, 1965), pp. 40-41.
9. D.M. Thomas, *Memories & Hallucinations* (Abacus, 1989), p. 47.
10. Bran Nicol, *D.M. Thomas* (Northcote House Publishers, 2004), p. 8.
11. See, for example, Bakhtin's discussion of the 'autonomous hero'. See also Gary Saul Morson, *Narrative and Freedom: the Shadows of Time* (Yale U.P., 1994).
12. Charlotte Brontë, 'Editor's Preface', p. 38 and p. 41.
13. I am reminded here of Cézanne: "The Landscape becomes reflective, human and thinks itself though me. I make it an object, let it project itself and endure within my painting . . . I become the subjective consciousness of the landscape, and my painting becomes its objective consciousness." (See 'David Bohm, Paul Cézanne and Creativity', F. David Peat, Pari Center website, www.paricenter.com)
14. Lewis Hyde, *The Gift: Imagination and the Erotic Life of Property* (Vintage, 1999), p. 167.
15. Emily Brontë, Chapter 9, *Wuthering Heights.*
16. Source now unknown; citation noted in my journal
17. Sulston's experience is an interesting echo of one of Einstein's boyhood moments of inspiration. According to Michio Kaku, Einstein recalled that he read 'with breathless attention' Aaron Bernstein's *Popular Books on Natural Science* in which the author asked children to visualise riding alongside an electrical signal inside a telegraph wire. (Michio Kaku, 'The Power of Staring', *New Scientist*, 16 April 2005), p. 48. Here, too, there is a suggestion that Einstein was willing, imaginatively, to take on new form as he tried to understand the nature of the electrical signal.
18. Lewis Hyde, op. cit.
19. From "The Foundations of Science" by Henri Poincaré, first published in Paris in 1908.
20. F.A. Kekulé/John Carey (ed), 'Chains and Rings: Kekulé's Dreams', *The Faber Book of Science* (Faber, 1995),p. 137.
21. ibid, p. 138.

22. ibid, p. 138.
23. Alan Lightman, 'Moments of Truth', *New Scientist* (19 November 2005), p. 36.
24. ibid, p. 36.
25. Source now unknown; citation noted in my journal.
26. Michio Kaku, *New Scientist,* op.cit., p.48.
27. See F. David Peat, 'Wolfgang Pauli: Resurrection of Spirit in the World', www.paricenter.com .
28. George Johnson, *Fire in the Mind: Science, Faith, and the Search for Order* (Vintage, 1996), p. 146.
29. Mark Doty, *My Alexandria* (Jonathan Cape, 1995), pp. 44-46.
30. Wim Wenders is by no means alone in his vision of beautiful dross. Take J.M.W. Turner, for example, and his remarkable paintings of the radiant haze of pollution over the Thames in the 1830s.
31. Ted Hughes (trans.), 'Introduction', *Tales from Ovid* (Faber, 1997).
32. Henry Corbin, *Avicenna and the Visionary Recital,* (Princeton U.P., 1960).
33. I'm reminded here of William Blake's view 'in which the visionary sees the whole visible universe transfigured, because he has "put on Imagination uncorrupt". In this state of lucidity, symbol and reality, Nature and Imagination, are seen to be one… "I knew that this world is a world of imagination and vision." See Evelyn Underhill on Blake in *Mysticism: The Nature and Development of Spiritual Consciousness* (Oneworld, 1999).

The thoughts of artist and sculptor Antony Gormley also seem relevant. He comments on the influence of the French phenomenologist, Merleau-Ponty, on his art: 'Imagination, he believed, can have privileged access to the hidden dimensions of Being—to what, in his later writings, he calls 'the invisible'. This, he said, can be imagined but not seen; it is not non-existent but pre-exists the visible. Every 'visible' dimension of being . . . is correlatively connected to an invisible or imaginary dimension; he sought to establish the real and the imaginary as two separate but corresponding realms: separate on the level of ordinary being but corresponding on the level of fundamental Being. Such correspondence even applied to the human body—"It is not the utilizable, functional, prosaic body which explains man: on the contrary, the body is precisely human to the extent that it discovers its symbolic and poetic charge."' *Antony Gormley* (Phaidon, 1995).

Antony Gormley

Antony Gormley *was born in London, England in 1950. After attending Ampleforth College and Trinity College, Cambridge, he travelled to India and Sri Lanka in order to deepen his study of Buddhism. Following his return to England he studied at the Central School of Art, Goldsmiths College and the Slade School of Art. In 1994 Gormley won the prestigious Turner prize.*

In much of his work Gormley has used casts of his own body in ways that confront the context of the space in which they are located. One of the largest of these, The Angel of the North, *is located in Gateshead, England and stands 66 feet tall and with wings 178 feet wide.*

THE INNER DARKNESS OF THE BODY

Interview with Antony Gormley

I first met Gormley at a dinner given by the sculptor, Anish Kapoor, and have since visited him at his studio on many occasions. In 1999 Gormley joined myself and the physicist, Basil Hiley, in a conversation on the notion of pre-space in quantum theory. During the discussion we mentioned the mathematician, David Hilbert's, definition of algebra as "the relationship of relationships". The remark struck Gormley and resulted in a series of new works called *Quantum Cloud*. In 2007 Gormley had a major new exhibition at the Hayward Gallery, London entitled *Blind Light*.

The particular conversation below took place in his studio in south London on June 4, 1996. We began discussing the way in which creative work begins within the physical body, frequently in the form of muscular tensions. During the gestation of the work these tensions have to be contained within the body.

Antony Gormley: I believe this. It's like giving birth. This thing is held and given form though being registered in the body. There is an accommodation in the body of the thing we have to do.

F. David Peat: The body plays a big role in your work. When you are making a cast what exactly do you do in preparation?

AG: I try not to get too . . . This morning we made the case for *Angel*. The period of preparation was very important. I knew it had to be with the chest fully inflated. I had to concentrate very, very hard on keeping the vertical that goes into the ground, and on the idea of the extension. The idea of the front of the body being extended. And for the first time I used mirrors as a register. As a matter of fact I put my back out!

FDP: Part of this space is not external. It's an internal space that the body occupies and we don't see from the outside.

AG: It's the other side of appearance. I was reading a paper by Stephen Levension, director of a school of anthropology. He wrote a paper on Kant turning up to a tribe of Navahos.

There are two theories about space—Newtonian absolute and independent. Then Leibniz who said it is relationship. Kant said both may be true but it's only experienced subjectively and our whole way of thinking about left, right, front, etc., is in relation to our bodies and we project that onto space.

I started thinking that all of that can be inverted. Our bodies are created by the potential of three dimensions. We become registers of an idea about space that is only experienced by the body. The space I'm interested in, and try to enter, is adimensional. It doesn't have this quality of dimension and it makes no sense to say "in front, behind, left, right". You lose all sense of those kinds of coordinates.

That Enlightenment idea of understanding the principles by which life is sustained has little to do with the space I'm interested in. Its a kind of . . . a darkness without fear. It may contain the possibility of evil but in some curious way, because you have entered voluntarily, the experience of it is about potential and power. Just as the spatial coordinates we use to make sense of the outside have to be left behind so too do moral coordinates.

I was once talking about the darkness of the body and someone said, "Oh, you mean evil". But no, I mean that darkness we carry with us always that is neither evil nor good but is the space of consciousness within the body.

FDP: Jung spoke about the shadow

AG: For him it's the other side of animus, the negative side of the projected personality. I don't think I read the darkness of the body in psychological terms either but in phenomenological terms. It is a place that has no dimensions.

I was very, very lucky as a child to have these powerful experiences of the space of the body because I was always sent off to go to sleep in these light-filled rooms, with a balcony. I did not need to sleep but would lie there and explore this space which was both incredibly claustrophobic and tiny but then began to expand and expand.

FDP: Hmm. So what of Anish Kapoor's voids?

AG: When I saw them first I thought, "I don't need to work anymore". He's made a phenomenal demonstration of the thing I wanted everyone to intuit without wanting to show it. I was as affected as everyone else. I thought they were breathtaking—the unlimited within the limited. I felt he has materialized the thing implicit in all my work.

Since then I realized that the thing I have to offer is this reflexivity that is not about. . . . The problem about Anish's work and James Turrell is that it becomes the demonstration of something like a mystical experience produced by tricks of light or absence of light. I would rather the hidden remained hidden. The work makes an absolute division. It undermines the dominance of appearance, not puncturing it.

The degree to which my work is unsatisfactory as a representation is the degree to which it is asking you to look for something in yourself that can empathize with the inner space of the work which is not an object, i.e. the degree as a representation because we look . . . , well, "the body in Western art". We look for beauty and a certain kind of likeness. My work doesn't give likeness, or beauty in any understood way. It really presents the body as a condition, not as a given identity. It should become a catalyst.

I want the circulatory of involvement. Look at the work, ask, "What is it doing here?" "How is it in space?" "What is its dialogue with space?" "Is there an interior/exterior tension?" Then reflectively you ask yourself the same thing, "What is the relationship of my interior?"

The impossible thing I'm trying to do is accept that we live in the world of the visible but make it unsatisfactory enough that behind the visible is some other kind of potential that does not exist in the sculpture but exists in you the viewer.

But maybe all this is a bit too airy-fairy.

FDP: With some sculpture you walk around and explore. Some push you into a position in space. Your work does put me into a position.

AG: I would like it to do both. But all of the work is relatively axially. The space is confrontational, and then you circle and navigate and then come back to one-to-one a kind of registration.

FDP: Last time I visited your studio you had several pieces around the walls. It was like an energy field. The internal space creates an external space of relationship.

AG: In the past I tried to deal with this by replacing the surface structure of skin and hair, etc. By a structure oriented to vertical and horizontal axes, a sort of matrix. Implied into this is the body of the planet.

FDP: That response to mass, I feel it in the solar plexus. It is kinesthetic.

AG: The way in which the work affects you somatically is strongest when things are off balance or standing on the wall, i.e. levers the room out of its repose, gives the idea is everything moving and nothing is fixed.

FDP: Does this also subvert the natural reaction to look for likeness?

AG: Anything that makes the viewer more conscious of his own weight, movement in space, his center of gravity while in constant motion. You are always falling when walking. That idea of human motion being precarious. I like the work to be still, silent and fixed but often tipped very close to be off its center of gravity. You think twice about taking for granted your own relationship to it. It is to encourage a reflexivity that is not about looking.

FDP: A clown with big boots is very disturbing.

AG: That's a very, very good image. Its funny because its absurd. The proportions are not as they should be. Yet in some curious way we all do those things mentally—in dreams or states of yearning. That's what made me make works like—the extended arms. It's a very, very common experience.

I remember the last time I had serious dentistry. I was on my bike and had the sensation I was rising until I was 20 ft. It was a clear sensation. I was in an extended body. I've had that feeling in other places as well. I wanted to give those natural sensations . . . the life of the imagination.

The sensate information from the matrix of the body is not always in sync with the . . . for example, Leonardo's inscribed body is an expression of Platonic absolutes. But our experience is not like that. Inside us there is always something else being born. We have bodies that are very good, provisional habitations for the spirit. We use them and through our time in the mind and the body we are making room or creating another kind of being. Those experiences of extension from the body are signs of the potentiality of that process.

Then there's that idea of who we are and what we look like. Your physiognomy belongs to me more than to you because I'm looking. The world of appearances is a shared communication. Where we derive energy in order to take part in the shared world of appearances is from the other side of them. I want to turn things round, or make you feel maybe there is somewhere outside the outside, or there are areas of experience that are independent of the functional side of the personal. I don't know whether this darkness is really collective.

Part of me would like to believe in the Theravada Buddhist tradition—that you can transmit love as a vibration which is independent of any object. It radiates out. It is registered in that space of the darkness of the body, rather than in the other world, the world of daily life, external appearances. I think that makes me feel there is a kind of collective experience of the inner space of the body. It's dangerous to think that kind of universal . . . everyone has a different relationship with the internal body.

FDP: Do you see that realization in the art of other cultures?

AG: It's in the wonderful Khmer heads. All of those south east Asian Buddhist sculptures have it. You find it occasionally in Western art. Rodin's *Age of Bronze* has it. He had it for one moment then lost it. He pushed. He was accused of casting it from life. He had a Belgian soldier who posed for a year and a half and nearly went mad. He is looking up and is sort of holding a spear but isn't. His eyes are closed and it becomes this internal moment of realization. All the attributes have been taken away that identified it with the 19th-centuryallegorical school and it was replaced by an idea about internal space.

FDP: The idea of history has gone.

AG: That is important to me. It's why the darkness of the body is important.

It's completely non conditional. People in the West are extremely frightened about this. It is only accessible through direct physical relationship and it's not put there or contextualized. It seems to many to be a denial of the whole positivistic and progressive idea about Western civilization. There has been an enormous resistance to my work because it is ahistorical.

In a way I'm accused of all things I'm not doing—i.e. looking for an ideal body. But I say look again. It's far from ideal. It's the body I am born with. Then they say, "Why is it generic?" And I say, "That's a function of making a case for something." It's inside that carries the index of the particular, the outside is just the brick.

(At this point the recording tape was changed. During this period Gormley referred to Mark Rothko whose work he felt was rather refined.)

AG: . . . I'm here for a while. How can I make an account for the world? Rothko is at the tail end of the sublime and is very refined. He feels he's started from first principles and there is nothing to be refined about. Trying to make distance count for something.

I'm aware of history of art but the thing is to be as direct as possible. I love Brancusi but it's the opposite, someone honing away, an abstraction that comes from purification of form which ends up being about reflected light and surfaces. I'm going the other way. If Brancusi and Cézanne used light to supercede distance, my ambition is to deal with darkness and the distance is . . . I don't know if I'm interested in distance at all. I want people to feel they're inside the work.

FDP: Brancusi and Cézanne draw a great deal on vision. Their work demands the eye but in your work these seems to be a great deal for a blind person.

AG: I think I'm part of a critique of the visual *per se*—that idea that you may use the visual sense as a basis. But it has to be verified by other forms that are to do with feeling—awareness of density. The stimulus, the way you may feel the density comes from the vision and you respond physically.

FDP: I think there is also a synesthesia involved in the way I respond to your work—to the lead or the iron. I have a certain taste in my mouth when I look at lead, I can feel its malleability.

AG: Yes, and when I use iron I'm aware it is an earth material. It has a strong gravity. It has a relationship with the liquid core of the earth. I think humans do this well, this recognition. We can't get away from it. But I don't want to deny vision. It is a primary sense. But what do you use it for? For part of a wider spectrum? I'd like to feel that I'm setting up a landscape, not just to look at but to walk through and become aware of the different fields.

FDP: I was listening to a talk by the film director, Ken Russell, who has recently been making radio plays. He says that vision can be shielded, that we can protect ourselves from shock, but that sound is more immediate and direct. So maybe some of these sensations you are talking about are internal and experienced in a more direct way while vision remains external and projected.

AG: Yes, I like that a lot. Sound being in some way something you are immersed in. It need not have an identifiable source. The sound of water in an underground cave, you are in that sound and the sound is in you. I'd like to feel the sculpture does that. It's being within being. It's a small and inert catalyst, a bit of matter used to catalyze your sense of being immersed in light and matter. Through the work you may become aware of the breath passing though the channels of your nose, or the weight running through your knees. Aware of the world that you inhabit and your aliveness within.

Its odd that I feel certain sculpture can do this so well, because it is the most difficult medium of all. It's so dumb, such a stumbling block for most people. Its a bit of the material world that causes you to look at it and where it is and how it is, in a way that also causes you to look at yourself.

I don't know. I think art in many ways is being hijacked by a limited idea. You see in those societies that don't want to be literate, they transfer their shared experiences in other ways. It's perfectly natural to paint the body and not so much decorate the shared world but make it responsive to thought. In *Field* and in *Angel of the North* it's about trying, in a way, to liberate art from this very, very limited world of the art gallery where things are so clinically and specifically contextualized.

[Note: *Field* consists of a very large number of small terracotta figures that totally fill the room of a gallery to confront the viewer. Gormley has made a number of versions of *Field*, in each case using ordinary people from the local community to mould the clay figures.]

FDP: Satish Kumar, from Schumacher College has a story about his mother weaving a shawl for his sister. The sister found it so beautiful that she told her mother it was too beautiful to wear and she would hang it on the wall. The mother said, "When you put beautiful things on a wall, you put ugly things on your body." Kumar feels that our society is behaving the same way, the more we collect things in galleries the uglier our daily world becomes.

AG: Is so weird. It's the trophy thing. Somehow having an animal's head on the wall or a souvenir is more important than having gone, or having seen an animal alive. Things should be quite natural.

FDP: There is a similar issue with much Native American art. Iroquois grave markers are supposed to remain outside until they rot and disappear. This is a big problem for museums who want to conserve such objects.

AG: I like the idea that I'm temporarily borrowing, like I do with my body, from the planet's matter/energy and chain of being. I borrow a few bits of material that I try to shape. In the same way I try to shape my life, or my life ties to shape me. Then what happens to the art does not really matter. I really like the idea that *Field* belongs back in the earth. It is to be buried or to melt back into the earth. The iron is strong but it is not protected from oxidation. I'd like one to be in a tank of water that would disappear. It's a temporary record of a moment in life. It may extend the image of that life but it has to go back into the cycle of things.

FDP: But there is also an element of accident involved. Some things are projected by nature, like cave paintings. Maybe you could put it under the sea so coral grows round it.

AG: And it becomes somebody else's raw material. I like that very much.

Pari Publishing is an independent publishing company, based in a medieval Italian village. Our books appeal to a broad readership and focus on innovative ideas and approaches from new and established authors who are experts in their fields. We publish books in the areas of science, society, psychology, and the arts.

Our books are available at all good bookstores or online at
www.paripublishing.com

If you would like to add your name to our email list to receive information about our forthcoming titles and our online newsletter please contact us at **newsletter@paripublishing.com**

Visit us at **www.paripublishing.com**

Pari Publishing

Pari Publishing Sas
Via Tozzi, 7
58040 Pari (GR)
Italy

Email: info@paripublishing.com